Weber's Planetary Model of the Atom

Figure 0.1:
Wilhelm Eduard Weber (1804–1891)
Foto: Gudrun Wolfschmidt in der Sternwarte in Göttingen

Nuncius Hamburgensis
Beiträge zur Geschichte der Naturwissenschaften
Band 19

Andre Koch Torres Assis, Karl Heinrich Wiederkehr
and Gudrun Wolfschmidt

Weber's Planetary Model of the Atom

Ed. by Gudrun Wolfschmidt

Hamburg: tredition science 2011

Nuncius Hamburgensis

Beiträge zur Geschichte der Naturwissenschaften

Hg. von Gudrun Wolfschmidt,
Geschichte der Naturwissenschaften, Mathematik und Technik,
Universität Hamburg – ISSN 1610-6164

*Diese Reihe „Nuncius Hamburgensis"
wird gefördert von der Hans Schimank-Gedächtnisstiftung.
Dieser Titel wurde inspiriert von „Sidereus Nuncius"
und von „Wandsbeker Bote".*

Andre Koch Torres Assis, Karl Heinrich Wiederkehr and Gudrun Wolfschmidt:
Weber's Planetary Model of the Atom. Ed. by Gudrun Wolfschmidt.
Nuncius Hamburgensis – Beiträge zur Geschichte
der Naturwissenschaften, Band 19. Hamburg: tredition science 2011.

*Abbildung auf dem Cover vorne und Titelblatt: Wilhelm Weber
(Kohlrausch, F. (Oswalds Klassiker Nr. 142) 1904, Frontispiz)*
*Frontispiz: Wilhelm Weber (1804–1891)
(Feyerabend 1933, nach S. 38, Abb. 17)*
*Abbildung auf dem Cover hinten:
Göttingen, Blick vom Wall auf die Sternwarte, um 1835,
Lithographie von Friedrich Besemann (1796–1854) – (Wiki)*

Geschichte der Naturwissenschaften,
Mathematik und Technik der Universität Hamburg
Bundesstraße 55 – Geomatikum, D-20146 Hamburg
http://www.math.uni-hamburg.de/spag/ign/w.htm

Dieser Band wurde gefördert von der Schimank-Stiftung.

Das Werk, einschließlich aller seiner Teile, ist urheberrechtlich geschützt. Jede Verwertung ist ohne Zustimmung des Verlages und des Autors unzulässig. Dies gilt insbesondere für Vervielfältigungen, Übersetzungen, Mikroverfilmungen sowie die Einspeicherung und Verarbeitung in elektronischen Systemen.

ISBN 978-3-8424-0241-6 – © 2011 Gudrun Wolfschmidt. Printed in Germany.

Contents

Preface:
Weber's Planetary Model of the Atom
Gudrun Wolfschmidt (Hamburg)   8

Einleitung: Wilhelm Webers Planetenmodell
Karl Heinrich Wiederkehr (Hamburg)   11

1 Weber's Planetary Model of the Atom
Andre Koch Torres Assis (Campinas, SP, Brazil) and Karl Heinrich Wiederkehr (Hamburg)   17
  1.1 Introduction 17
  1.2 Weber's Atomism 18
  1.3 The Nature of the Galvanic Current 19
      1.3.1 The Electric Current According to Ørsted 20
      1.3.2 The Electric Current According to Ampère 22
      1.3.3 The Electric Current According to Faraday 24
      1.3.4 The Electric Current According to Maxwell 28
      1.3.5 Weber's Initial Conception of an Electric Current . . . 31
  1.4 The Nature of the Molecular Currents 34
      1.4.1 The Molecular Currents According to Ampère 34
      1.4.2 On the Origins of the Resistance of Conductors According to Weber 36
      1.4.3 The Molecular Currents According to Weber 41
      1.4.4 Further Developments of Weber's Conception about Ampère's Molecular Currents 44
  1.5 The Evolution of Weber's Conception of an Electric Current: From a Double Current to a Simple Current 48
  1.6 The Motion of Two Charged Particles Interacting According to Weber's Force Law 53
      1.6.1 Weber's Force and Potential Energy 53
      1.6.2 Weber's Introduction of an Inertial Mass for the Electric Fluids 54

      1.6.3   Weber's Equation of Motion and His Critical Distance . 57
      1.6.4   Motion of Two Particles of the Same Kind 59
      1.6.5   Motion of Two Dissimilar Electrical Particles 61
1.7   Weber's Speculations about the Conduction of Electricity and Heat in Conductors 62
1.8   Weber's Speculations about the Conduction of Heat in Insulators 67
1.9   Optical Properties of Weber's Planetary Model of the Atom . . 70
1.10 Weber's Mature Planetary Model of the Atom and the Periodic System of the Elements 76
      1.10.1 Deriving a Gravitational Force Law from Weber's Electric Force Law 76
      1.10.2 The Manifold of Ponderable Bodies 78
      1.10.3 The Periodic Table of the Chemical Elements 81
      1.10.4 Application to Chemical Bondings 85
      1.10.5 Open Topics 87
1.11 Final Considerations 88
1.12 Acknowledgments 90
Assis and Wiederkehr: Bibliography 91

2 Vorstellungen von der elektrischen Leitung und Entwicklung der Elektronentheorie der Metalle von Riecke, Drude, Lorentz bis Sommerfeld
Karl Heinrich Wiederkehr und Gudrun Wolfschmidt (Hamburg)   103
  2.1   Einleitung 103
  2.2   Michael Faraday (1791–1867) und André Marie Ampère (1775–1836) 104
  2.3   James Clerk Maxwell und die „Natur des elektrischen Stroms". 106
  2.4   Gustav Theodor Fechner (1801–1887) und Wilhelm Weber (1804–1891) 108
  2.5   Die klassische Elektronentheorie der Metalle 116
  2.6   Eduard Riecke (1845–1915) 117
  2.7   Paul Drude (1863–1906) 125
  2.8   Ludwig Lorenz (1829–1891), dänischer Physiker 127
  2.9   Hendrik Antoon Lorentz (1853–1928) 128
  2.10 Niels Bohr (1885–1962) 130
  2.11 Krise der klassischen Elektronentheorie der Metalle 132
  2.12 Stewart-Tolman-Versuch – Richard Chace Tolman (1881–1948) und Thomas Dale Stewart (1890–1958) 133
  2.13 Arnold Sommerfeld (1868–1951) 134
  2.14 Literatur 142

3 Works by Assis related to Weber's Law
　　Applied to Electromagnetism and Gravitation
　　Andre Koch Torres Assis (Campinas, SP, Brazil)   151
　　3.1   Books 151
　　3.2   Papers 152
　　4.1   Zeittafel zu Wilhelm Weber (1795–1878) 164

4 Arbeiten zu Gauss und Weber von K. H. Wiederkehr
　　Karl Heinrich Wiederkehr (Hamburg)   165

Autors   167

List of Figures   171

Nuncius Hamburgensis   173

Index   179

Preface:
Weber's Planetary Model of the Atom

It is well known that Wilhelm Eduard Weber (1804–1891) was hired as professor in Göttingen with the recommendation of Carl Friedrich Gauß (1777–1855). The result of their cooperation was the first electromagnetic telegraph in 1833 and the foundation of a geomagnetic observatory in Göttingen. The "Magnetic Association" (*Magnetischer Verein*) became very influential on an international level with about 50 observatories in all five continents. An important result of their cooperation was the publication of the *Atlas des Erdmagnetismus* by Gauß and Weber in 1840. About this topic an exhibition was organized by Gudrun Wolfschmidt and Karl Heinrich Wiederkehr in the State and University Library Hamburg in 2005 and a book was published.[1]

Wilhelm Weber started with research in the whole field of magnetism and electrodynamics. Weber's capacity in making precision experiments is famous leading to the absolute system of units, the fundament of our SI-units. The results were published in 1864 as *Electrodynamic Proportional Measures*.

PD Dr. Karl Heinrich Wiederkehr is a pioneer in the field of research concerning Wilhelm Weber and the development of electrodynamics. His dissertation *"Wilhelm Webers Stellung in der Entwicklung der Elektrizitätslehre"* (1960) and the biography of *Wilhelm Eduard Weber – Erforscher der Wellenbewegung und der Elektrizität 1804–1891* (Vol. 32 of the *Reihe Große Naturforscher*), published in 1967, were well acknowledged in the scientific community (cf. the review in Isis and in other scientific journals).

During his time as a teacher at a secondary school *Gymnasium* (most recently as *Studiendirektor* and *Koordinator der Oberstufe*) Karl Heinrich Wiederkehr worked as a freelancer after the founding of the *Institute for History of Science and Technology* (1960) without taking a reduction in teaching time. Together with Bernhard Sticker, director of the Institute, he held several talks, especially at the Institute of Teacher Education (*Landesinstitut für Lehrerbildung*

---

1 Wolfschmidt, Gudrun (Hg.): Vom Magnetismus zur Elektrodynamik. Herausgegeben anläßlich des 200. Geburtstages von Wilhelm Weber (1804–1891) und des 150. Todestages von Carl Friedrich Gauß (1777–1855). Katalog zur Ausstellung in der Staatsbibliothek Hamburg, 3. März bis 2. April 2005. Hamburg: Institut für Geschichte der Naturwissenschaften 2005.

Figure 0.2:
Gauss-Weber-Exhibition *Vom Magnetismus zur Elektrodynamik*
of the Institute for History of Science and Technology, Hamburg University,
organized by Gudrun Wolfschmidt and Karl Heinrich Wiederkehr,
in the State and University Library Hamburg in 2005
Photo: Gudrun Wolfschmidt

und Schulentwicklung) in Hamburg. For the North German radio broadcasting *Norddeutscher Rundfunk* (NDR 3), he wrote text books for the series *Classical Experiments of Physics* (1969) and he organized and designed experiments. In the encyclopaedia *Große Naturwissenschaftler* (1970), ed. by Fritz Krafft and Adolf Meyer Abich, he wrote more than 80 short biographies.

In 1974 Karl Heinrich Wiederkehr finished his habilitation with the title: *René-Just Haüys Vorstellungen vom Kristallbau und einer chemischen Atomistik*. It was published in four parts in *Centaurus* in 1977 and 1978. This work was reviewed and quoted in the scientific litersture (eg. J. J. Burckhardt: *Die Symmetrie der Kristalle* (Birkhäuser 1988). The book *Physics and Geophysics with Historical Case Studies* (1997), ed. Wilfried Schroeder, presented to his 75th birthday, contains a list of his numerous scientific papers in various fields of physics and geophysics (you find also in this book a list of Wiederkehr's publications, p. 165).

After his retirement, Dr. Wiederkehr worked again at a larger extent in the Institute as *Privatdozent*. In cooperation with Gudrun Wolfschmidt he produced some exhibitions *From Magnetism to Electrodynamics* (2005), *Von Hertz zum Handy* (the development of communication)² (2007–2008) and *History of Navigation* (2008–2010) and contributed considerably to the publications connected to the exhibitions.³

Together with Andre Koch Torres Assis from Brazil, another important scholar in the field of Wilhelm Weber (cf. a list of his publications can be found also in this book, p. 151), Dr. Wiederkehr cooperated in the years 2001, 2002 and 2009, during his stay in our Institute in Hamburg. Weare glad that this further cooperation with Assis was possible, payed by the Humboldt foundation. In this book *Weber's Planetary Model of the Atom* is discussed by Assis in detail.

The next article in this book by Wiederkehr and Wolfschmidt discuss the further development in the 20th century, especially the metal electron theory. Around 1920 atomic physics was developed quickly on the basis of Sommerfeld's influential book *Atombau und Spektrallinien* (Braunschweig 1919). The centers of development were Munich, Göttingen, Zürich and Leipzig. In the middle of the 1920s the theory of quantum mechanics was created (Heisenberg / Born / Jordan, Schrödinger and Dirac). At the end of the 1920s very quickly a number of applications followed with Pauli and Sommerfeld as the leading persons.⁴ Here the school of Sommerfeld contributed considerably.⁵ In 1926 quantum statistics was introduced; the results were compiled by Bethe in his famous article *Elektronentheorie der Metalle* (1933). The further development of the metal electron theory and the beginning of solid state physics is discussed by Eckart/Schubert (1986/1990) and by Hoddeson et al.: *Out of the Crystal Maze* (New York 1990).

Gudrun Wolfschmidt

---

2 Wolfschmidt, Gudrun (Hg.): Von Hertz zum Handy – Entwicklung der Kommunikation. Norderstedt: Books on Demand (Nuncius Hamburgensis; Band 6) 2007. Wolfschmidt, Gudrun (ed.): Heinrich Hertz (1857–1894) and the Development of Communication. Proceedings of the International Symposium in Hamburg, October 8-12, 2007. Norderstedt: Books on Demand (Nuncius Hamburgensis; Bd. 10) 2008.

3 Wolfschmidt, Gudrun (Hg.): „Navigare necesse est" – Geschichte der Navigation. Norderstedt: Books on Demand (Nuncius Hamburgensis; Band 14) 2008. Wolfschmidt, Gudrun: „Sterne weisen den Weg"' – Geschichte der Navigation. Norderstedt: Books on Demand (Nuncius Hamburgensis; Band 15) 2009.

4 Casimir, Hendrik B. G.: Pauli and the Theory of Solid State. In: Theoretical Physics in the Twentietrh Century. New York 1960.

5 Eckert, Michael: Die Atomphysiker. Eine Geschichte der theoretischen Physik am Beispiel der Sommerfeldschule. Braunschweig: Vieweg 1993.

Einleitung: Wilhelm Webers Planetenmodell

Während meiner Studienzeit an der Universität in Hamburg vom Winter 1945 bis zum Sommer 1949 in Physik, Mathematik, Chemie und Philosophie hörte ich regelmäßig Vorlesungen bei Prof. Dr. Adolf Meyer-Abich (1895–1971) und beim Honorar-Prof. Dr. Hans Schimank (1888–1979).

Adolf Meyer-Abich war Inhaber des Lehrstuhls für das Fach Philosophie und Geschichte der Naturwissenschaften, Hans Schimank hatte einen besoldeten Lehrauftrag für die Geschichte der Physik, Chemie und Technik. Daneben leitete er das Technische Vorlesungswesen an der Ingenieurschule am Berliner Tor, heute Hochschule für Angewandte Wissenschaften. Den Vorlesungen Hans Schimank zu folgen war ein Genuss, sprachlich wie inhaltlich. Zur Veranschaulichung projizierte er viele Bilder an die Wand und reichte aus seiner eigenen Bibliothek zeitgenössische Literatur und Originalwerke der gerade behandelten Forscher herum. Meyer-Abich spannte mehr einen philosophischen Bogen über seine Themen, Schwerpunkt war bei ihm die Biologie und Nachbargebiete. Mit seiner holistischen Sicht erweiterte er den Horizont seiner Zuhörer.

Im Januar 1950 legte ich mein Staatsexamen für das Lehramt an Gymnasien in den oben genannten Fächern ab und war im gleichen Jahr bereits 1950 im Hamburger Schuldienst.

Mein Studium war oft stressig und hektisch gewesen. Ich wollte möglichst schnell in den Beruf kommen. Ruhepunkte und Mußestunden waren für mich die oben genannten Vorlesungen zur Geschichte der Naturwissenschaften gewesen. Während des Schuldienstes Mitte der 50er Jahre ließ ich mir von Hans Schimank, damals Nestor der Naturwissenschaftsgeschichte in Deutschland genannt, das Thema für eine Dissertation geben mit dem Titel *„Über Wilhelm Webers Stellung in der Entwicklung der Elektrizitätslehre"*. Hans Schimank hatte in mir die Liebe zur Geschichte der Naturwissenschaft geweckt und den Drang entfacht, den Wegen des menschlichen Geistes nachzuspüren. Ich war der erste Doktorand von Hans Schimank. Alle meine wissenschaftlichen Arbeiten habe ich angefertigt – auch meine spätere Habilitationsschrift – ohne irgendeine Ermäßigung in der von der Behörde vorgeschriebenen Zahl von Unterrichtsstunden.

Über den großen Physiker Wilhelm Weber fand ich in der Literatur verschiedene, oft sich widersprechende Ansichten vor, und ich entschloss mich, sein umfangreiches Werk durchzuarbeiten. Und dies nahm ein paar Jahre in Anspruch.⁶

Um an einen eventuell vorhandenen Nachlass zu kommen, wandte ich mich an Nachkommen Wilhelm Webers und seiner zwei Brüder, die ebenfalls bekannte und auch berühmte Professoren in der Physiologie und Anatomie waren. Leider musste ich erfahren, dass nur noch Weniges nach dem Zweiten Weltkrieg vorhanden war. Persönliche Briefe von und an W. Weber, insbesondere die Briefe an Gauß, fand ich in Göttingen. Es war W. Webers Herzenswunsch gewesen, sich in der Nähe von Gauß wissenschaftlich vervollkommnen zu können. Der zu seiner Zeit schon renommierte Akustiker wurde von Gauß in die Gebiete des Magnetismus und der Elektrizität eingeführt. Von Gauß und Weber wurde der „Göttinger *Magnetische Verein*" ins Leben gerufen. Der *Magnetische Verein* war leuchtendes Vorbild für die spätere Kooperation im Ersten Polarjahr 1883/84. Die vielversprechende Zusammenarbeit zwischen Gauß und Weber wurde damals durch die Entlassung Webers jäh unterbrochen, W. Weber gehörte zu den *Göttinger Sieben*, Vorboten für die Revolution 1848/49 in Deutschland.

In der zweiten Hälfte der 50er Jahre legte ich meinen Dissertationsentwurf Hans Schimank vor. Fast ein ganzes Semester lange wurde in den von Schimank abgehaltenen Übungen und Besprechungen zur Geschichte der Naturwissenschaften über meine Arbeit diskutiert. 1960 wurde in Hamburg das *Institut für Geschichte der Naturwissenschaften, Mathematik und Technik* gegründet, als Leiter des Instituts berief man Prof. Bernhard Sticker, und unter ihm promovierte ich im Januar 1962 mit der von Hans Schimank gestellten Arbeit. Die Dissertation „*Wilhelm Webers Stellung in der Entwicklung der Elektrizitätslehre*", Hamburg 1960 wurde recht positiv und anerkennend besprochen, auch in maßgeblichen internationalen Zeitschriften. Meine Recherchen über W. Weber hatten sich natürlich nicht nur auf Arbeiten über Magnetismus und Elektrizität beschränkt. 1967 erschien von mir eine Biografie Wilhelm Webers als Band 32 der Reihe „Große Naturforscher". Darin wurde das Gesamtwerk Wilhelm Webers gewürdigt; der Band fand ebenfalls eine positive Aufnahme. Ein ganzer Kranz von Abhandlungen behandelte spezielle Themen, so „Aus der Geschichte des Göttinger Magnetischen Vereins und seine Resultate" (1964) und „Hamburgs patriotische Bürger und die Göttinger Sieben" (1964).

Nach der Öffnung der Berliner Mauer 1989 schenkte man in Lutherstadt-Wittenberg, Halle und Leipzig Wilhelm Weber und seinen beiden Brüdern

---

6 Siehe hierzu Wiederkehr, K. H.: Mein Weg zur Geschichte der Naturwissenschaften – Anstöße, Begegnungen und Arbeiten. In: Schröder, Wilfried (Hg.): Wege zur Wissenschaft, Pathways to Science. Bremen-Rönnebeck, Potsdam: Science Edition 2001, S. 257–264.

Abbildung 0.3:
Die drei Weber-Brüder:
Ernst Heinrich, Wilhelm Eduard und Eduard Friedrich

Ernst Heinrich und Eduard Friedrich mehr Aufmerksamkeit. Der ältere Bruder, Ernst Heinrich Weber (1795–1878), aber auch Wilhelms jüngerer Bruder, Eduard Friedrich (1806–1871), gehörten damals zur Elite der Physiologen und Anatomen. Die beiden führten in die Physiologie exakte naturwissenschaftliche Methoden ein. In der Martin-Luther-Universität Halle-Wittenberg und in der Universität Leipzig fanden über die drei Weber-Brüder Symposien und andere Veranstaltungen statt und auf Einladungen hin nahm ich mit Vorträgen immer daran teil.

Auch im Ausland, in Brasilien, entfachte die fast vergessene ältere Elektrodynamik bei einem jungen Physiker, Andre Koch Torres-Assis, Professor an der Universität Campinas, großes Interesse. Die Arbeiten von André Marie Ampère und Wilhelm Weber faszinierten ihn. Sein Buch *„Weber's Electrodynamics"*, erschienen 1994, lässt Webers Werk in der Elektrizitätslehre in einem neuen Licht erscheinen. Dem Webersche Grundgesetz der elektrischen Wirkungen (1846) gibt er eine neue Fassung unter zu Hilfenahme von Vektoren. Die schwierigen Begriffe der relativen Geschwindigkeit und Beschleunigung werden so klarer und neben vielem Anderen kann Assis so aus dem Weberschen Gesetz die Formel für die Lorentzkraft herleiten. Assis zeigt auch, dass das Webersche Wechselwirkungsgesetz für elektrische Ladungen in beschleunigten Bezugssystemen erhalten bleibt, – im Gegensatz zu damaligen anderen Grundgesetzen der elektrischen Wirkungen. Neben zahlreichen Arbeiten zu dieser Thematik erschien 13 Jahre später von Assis und Julio Akashi Hernandes ein anderes wichtiges Werk *„The Electric Force of a Current"*. Es behandelt die von W. Weber geforderten Oberflächenladungen auf einem stromdurchflossenen Leiter und seine Folgerungen und räumt mit allzu schnell gefassten negativen Urteilen über das Webersche Gesetz auf. Dank der Alexander von Humboldt-Stiftung und dem Engagement von Frau Prof. Dr. Karin Reich konnte Herr Assis 2002 nach Deutschland kommen und am Institut für Geschichte der Naturwissenschaften, Mathematik und Technik in Hamburg mit mir ein Jahr zusammenarbeiten. Die gemeinsamen Arbeiten hatten die magnetischen und elektrischen absoluten Maßsysteme, die auf Gauß und Weber zurück gehen, zum Schwerpunkt. Die heutigen SI-Einheiten haben als Grundlage das absolute elektromagnetische Maßsystem. Im Kohlrausch-Weber Experiment (1855) wurde mit rein magnetischen und elektrischen Messungen die Lichtgeschwindigkeit gefunden. Für Maxwell war dies ein wichtiger Anstoß für seine elektromagnetische Lichttheorie gewesen.

Sieben Jahre später konnte Herr Assis dank der Humboldt-Stiftung und der Empfehlung von Frau Prof. Dr. Gudrun Wolfschmidt abermals in Hamburg ein Vierteljahr mit mir zusammenarbeiten. Das Ergebnis ist das hier vorliegende Buch. Ansatzpunkt für Webers Planetenmodell war in Anlehnung an A. M. Ampère seine Vorstellungen über fließende elektrische Teilchen in einem metallischen Leiter und die Amperesche Molekularstromhypothese. In meiner Dissertation (1960) war dazu schon einiges angeschnitten worden. In der vorliegenden Arbeit fand eine tiefere Analyse statt und Webers Gedanken wurde eingehender nachgespürt.

Verwunderlich bleibt, dass in der Physikgeschichte mit wenigen Ausnahmen (so bei Edmund Hoppe) über diese spekulative Seite kaum berichtet wird. Der Gedanke drängt sich auf, dass Hermann von Helmholtz (1821–1894), auch

„Reichskanzler" in der damaligen Physik genannt, – er wird auch gelegentlich als „Intimfeind" von Wilhelm Weber und seiner Brüder bezeichnet (so nach Herbert Hörz) – dank seines Einflusses und Ansehens mit im Spiele ist. Helmholtz und der mit ihm befreundete Emil Du Bois-Reymond (1818–1896) fühlten sich zu Beginn ihrer Karriere von den in der Physiologie dominierenden Weber-Brüdern nicht genug gewürdigt bzw. unterdrückt.[7] Ein enger Freund Wilhelm Webers, Johann Christian Poggendorff (1796–1877), hatte auch die Arbeit von Helmholtz „Über die Erhaltung der Kraft" (1847) für eine Aufnahme in die Annalen der Physik und Chemie „als zu spekulativ" abgelehnt. Helmholtz sah in seiner Arbeit das Webersche Gesetz als nicht vereinbar mit dem Satz von der Erhaltung der Energie an – zu Unrecht, wie Weber später nachwies. Webers Elektrodynamik wurde allzu schnell als überholte Fernwirkungstheorie abgetan, und mit ihr die später so fruchtbaren atomistischen Vorstellungen von der Elektrizität.

Webers Planetenmodell lieferte eine tragfähige Grundlage zur Deutung des Diamagnetismus. Nach der Entdeckung des Elektrons 1897 durch J. J. Thomson, wurden zahlreiche Atommodelle entwickelt –, mehrere von Thomson selbst. Die meisten Modelle verwarf man wieder relativ schnell, im Gegensatz dazu hielt Wilhelm Weber den einmal eingeschlagenen Weg konsequent bei. Stütze waren ihm hierbei eigene experimentelle Erfahrungen. Das vorliegende Buch will die Geschichte der Physik des 19. Jahrhunderts in Einigem ergänzen und Manches zurechtrücken. Mit Webers Entwicklung seines Planeten-Atommodells hängen eng auch die Anfänge einer Metallelektronik zusammen. Eduard Riecke nahm dann die Weberschen Ideen auf und entwickelte sie weiter. Aufgrund der Bedeutung der Entwicklung der Elektronentheorie wurde über dieses Thema ein eigener Artikel beigefügt. Im Anhang finden sich die Arbeiten der Autoren über Gauß und Weber.

Karl Heinrich Wiederkehr

---

7 Hörz, Herbert: Physiologie und Kultur in der zweiten Hälfte des 19. Jahrhunderts – Briefe an und von Hermann von Helmholtz. Rangsdorf: Basilisken Presse 1984. - Schröer, Heinz: Carl Ludwig. Stuttgart (Große Naturforscher; Band 33) 1967. - Dubois-Reymond, Estelle (Hg.): Zwei große Naturforscher des 19. Jahrhunderts – ein Briefwechsel zwischen Emil Du Bois-Reymond und Karl Ludwig. Leipzig: J. A. Barth 1927, S. 49–53. - Assis, K. T. und K. H. Wiederkehr: Weber quoting Maxwell. In: Mitteilungen der Gauss-Gesellschaft, Nr. 40, Göttingen 2003, S. 53–74, bes. 55 und 56.

Figure 1.1:
Wilhelm Eduard Weber (1804–1891) – 1835
(Lithographie im Besitz der Städtischen Altertumssammlung Göttingen)
Feyerabend 1933, nach S. 36, Abb. 16.

Weber's Planetary Model of the Atom

Andre Koch Torres Assis (Campinas, SP, Brazil) and
Karl Heinrich Wiederkehr (Hamburg)

1.1 Introduction

Wilhelm E. Weber (1804–1891) was one of the main scientists of the XIXth century. His complete works were published in 6 volumes between 1892 and 1894.[1] He wrote eight major Memoirs between 1846 and 1878 under the general title *Elektrodynamische Maassbestimmungen (Electrodynamic Measurements, Determination of Electrodynamic Measures or Electrodynamic Measure Determinations).*[2] The eighth Memoir was published only posthumously in his collected papers.

Three of these eight Memoirs have already been translated into English, namely, the first, *Determinations of electrodynamic measure: Concerning a universal law of electrical action;*[3] the sixth, *Electrodynamic measurements – Sixth Memoir, relating specially to the principle of the conservation of energy;*[4] and the eighth, *Determinations of electrodynamic measure: Particularly in respect to the connection of the fundamental laws of electricity with the law of gravitation.*[5] In 1848 it was published an abridged version of the first Memoir,[6] which has also been translated into English, *On the measurement of electrodynamic forces.*[7]

---

1 [Web92b, Web92a, Web93, Web94b, WW93, WW94].
2 [Web46, Web52b, Web52a, KW57, Web64, Web71, Web78, Web94a].
3 [weba].
4 [Web72].
5 [webb].
6 [Web48a].
7 [Web66c].

Here we follow the development of Weber's planetary model of the atom.

1.2 Weber's Atomism

In this work we present some very specific lines of reasoning which led Wilhelm Weber to his planetary model of the atom. They are related with the nature of the galvanic current, with the origin of the resistance of conductors, with Ampère's molecular currents, with Weber's fundamental force law of interaction between charged particles, and with the nature of magnetic and diamagnetic phenomena. Before going into these details, it seems important here to emphasize Weber's general conceptions about the nature of physical phenomena. During all his scientific career he presented a corpuscular or atomistic conception of nature. According to this idea, the universe has a granular constitution with a vacuum between these grains. That is, he assumed an atomic constitution of matter. Weber utilized different names to characterize these corpuscles, namely, atoms, molecules, particles etc. The interaction between these corpuscles can be thought as being effected through their collisions, or through forces acting at a distance between these particles. Weber considered the existence of an ether between these particles. But even this ether was thought by Weber in corpuscular terms, namely, as being composed of charged particles moving in space.

One of Weber's first works was a joint book with his older brother, Ernst Heinrich Weber (1795–1878), published in 1825, *Wave Theory Based on Experiment*.[8] In this work they dealt with waves in liquids and presented also applications to sound-waves and to light-waves. All these waves were supposed to propagate in a molecular medium.

Wilhelm Weber utilized the atomic hypothesis in all areas of physics in which he worked. He interpreted the phenomena in terms of an interaction between the smallest particles involved in the process. Through this interaction, these particles changed their locations and motions, creating all kinds of vibrations in a medium composed of these particles. He was convinced about the existence of discrete elementary particles. At the end of his life he tried to derive, at least qualitatively, the main physical phenomena based only on the existence of positive and negative charged particles interacting with one another according to his fundamental force law of 1846.

Weber's point of view was not shared by many physicists of the XIXth century. Instead of this atomistic hypothesis, these other scientists accepted a

---

8 [WW93].

dynamical conception of the world based on the existence of a continuum substance filling all space. Instead of a law of interaction between atoms or molecules, the dynamists were looking for an equation describing the flow of this continuous substance through space utilizing concepts like fields, lines of force, vortexes etc. Faraday and Maxwell are some of the most prominent scientists who advocated this line of reasoning in their works about electricity. We discuss their points of view in Sections 1.3.3 and 1.3.4.

Although Weber had always maintained an atomistic point of view, he certainly went more deeply into this matter due to the influence of his friend Gustav Theodor Fechner (1801–1887).[9] Fechner worked in many areas of physics, physiology, psychology and philosophy. He was one of the first scientists who recognized the significance of Ohm's work. He was a militant advocate of atomism. He wrote a famous book on this subject, *Ueber die physikalische und philosophische Atomenlehre* (Physical and Philosophical Conceptions of Atomism, or Physical and Philosophical Atomic Theory). The first edition was published in 1855 and the second one in 1864.[10] In this work he advanced this basic principle of atomism in order to explain many physical phenomena and laws of nature. He translated into German Biot's book *Précis élementaire de Physique expérimentale*. In separate chapters written by himself, Fechner presented the idea that the heat particles orbited around the atom like the planets around the Sun.[11] Later on Weber developed a similar model of the atom, but replacing the heat particles by electrical particles.

Fechner was educated at the University of Leipzig and later on was appointed its professor of physics. He resigned his position after contracting an eye disorder in 1839. Weber worked at the University of Leipzig from 1843 to 1849. During this period he had long discussions with Fechner about physics and philosophy. Both friends influenced one another.

1.3 The Nature of the Galvanic Current

Here we follow the conceptions of the electric current according to Ørsted, Ampère, Faraday, Maxwell and Weber. Ørsted, Ampère and Weber, in particular, supposed the electric current to be composed of a double flux of positive and negative charges moving in opposite directions with the same velocity relative to a current carrying wire. Weber later on considered the possibility of

---

[9] [Hei93] and [Hei04].
[10] [Fec64].
[11] [Bio28, Viertes Schaltcapitel: Ueber den wahrscheinlichen Grundzustand der Körper, p. 396–411, see especially p. 409–410].

each positive charge staying fixed with an atom or molecule composing the conductor, while the negative charge would orbit around it in elliptical orbits, in analogy with Kepler's planetary laws. With the application of an applied electromotive force, only the negative charges would move relative to the wire. This was one of the origins of his planetary model of the atom.

1.3.1 The Electric Current According to Ørsted

In 1800 Volta (1745–1827) (cf. Figure 2.2, S. 104) presented to the world his pile or electrical battery.[12] A source of power was created with this instrument which produced a relatively constant and large electric current around a metallic circuit connected to the extremities of this battery. The battery produced a small tension between its ends. Volta himself spoke of a "current of electricity" through the conductor connected to his pile.

Hans Christian Ørsted (1777–1851) was greatly influenced by the romantic culture of German Naturphilosophie. This was a philosophical tradition of German idealism in the XIXth century. According to these ideas, there was an unity in nature, with all phenomena being connected as a whole. Ørsted, in particular, looked for effects indicating the unity in different realms: heat, light, electricity, magnetism, chemistry etc. In 1820 he observed the deflection of a magnetic needle from its normal orientation along the terrestrial magnetic meridian when an electric current flowed along a nearby long straight conductor.[13] With this discovery, he opened a new field of research dealing with the interaction of electricity and magnetism. He created the name "electromagnetism" to characterize this new area.[14]

Instead of a "current of electricity," Ørsted utilized the expression "conflict of electricity." He defined it as follows:[15]

> The opposite ends of the galvanic battery were joined by a metallic wire, which, for shortness sake, we shall call the uniting conductor, or the uniting wire. To the effect which takes place in this conductor and in the surrounding space, we shall give the name of the conflict of electricity.

His explanation for the deviation of the magnetic needle from its normal orientation when close to a current carrying wire went as follows:[16]

---

12 [Vol00a] and [Vol00b].
13 [Oer20].
14 [Ørs98], [GG90, p. 920] and [GG91].
15 [Oer20].
16 [Oer20].

Figure 1.2:
Hans Christian Øersted (1777–1851)
Lenard 1930, p. 189.

We may now make a few observations towards explaining these phenomena. The electric conflict acts only on the magnetic particles of matter. All non-magnetic bodies appear penetrable by the electric conflict, while magnetic bodies, or rather their magnetic particles, resist the passage of this conflict. Hence they can be moved by the impetus of the contending powers.

It is sufficiently evident from the preceding facts that the electric conflict is not confined to the conductor, but dispersed pretty widely in the circumjacent space.

From the preceding facts we may likewise infer that this conflict performs circles; for without this condition it seems impossible that the

> *one part of the uniting wire, when placed below the magnetic pole, should drive it towards the east, and when placed above it towards the west; for it is the nature of a circle that the motions in opposite parts should have an opposite direction. Besides, a motion in circles, joined with a progressive motion, according to the length of the conductor, ought to form a conchoidal or spiral line; but this, unless I am mistaken, contributes nothing to explain the phenomena hitherto observed.*
>
> *All the effects on the north pole above-mentioned are easily understood by supposing that negative electricity moves in a spiral line bent towards the right, and propels the north pole, but does not act on the south pole. The effects on the south pole are explained in a similar manner, if we ascribe to positive electricity a contrary motion and power of acting on the south pole, but not upon the north. The agreement of this law with nature will be better seen by a repetition of the experiments than by a long explanation. The mode of judging of the experiments will be much facilitated if the course of the electricities in the uniting wire be pointed out by marks or figures.*

So, in essence, Ørsted believed that there was a double current of positive and negative electricities moving inside the wire in opposite directions. He ascribed a "contrary motion" to the positive and negative electricities not only inside the wire, but also outside it. From this expression we can infer that he believed these opposite electricities to move with velocities of the same magnitudes. Outside the wire there would also exist this double current, but now moving in a conchoidal or spiral line. It would be composed of a motion in circles, joined with a progressive motion, according to the length of the conductor.

1.3.2 The Electric Current According to Ampère

A.-M. Ampère (1775–1836) began his works on electrodynamics after Ørsted's discovery. However, he rejected Ørsted's idea of something like positive and negative electricities flowing in circles outside the wire. But he accepted the idea that inside a current carrying wire there was a double current of positive and negative electricities moving in opposite directions along the wire:[17]

> *[...] a double current thus results, the one positive electricity and the other negative electricity, moving in opposite senses from the*

---

17 [Amp20a, p. 64], [Far22, p. 112] and [Amp65b, p. 141].

Figure 1.3:
André Marie Ampère (1775–1836)
Lenard 1930, p. 197.

points where the electromotive action takes place to meet again in the part of the circuit opposite these points.

The same idea has been expressed a little later with similar words in the same work:[18]

Such are the differences which were known to exist between the effects produced by electricity in its two states which I have just described [that is, electric tension and electric current], the one being, if not a state of rest, at least one of slow motion due solely to the difficulty of isolating bodies in which electric tension occurs, the other

---
18 [Amp20a, p. 68–69] and [Amp65b, p. 144].

> being the double flow of positive and negative electricity along a continuous circuit of conducting bodies. In the conventional theory of electricity the two fluids of which it is thought to be constituted, are conceived to be perpetually separated in a part of the circuit and to be carried rapidly in contrary senses into another part of the circuit where they are continually re-uniting.

1.3.3 The Electric Current According to Faraday

This conception of the current as a flow of charged particles was not accepted by all scientists of the time. Michael Faraday (1791–1867) and James Clerk Maxwell (1831–1879), for instant, rejected it. In this Section we consider Faraday's point of view and in the next one Maxwell's conception.

Faraday began his works on electromagnetism after Ørsted's discovery of 1820. Initially he repeated some of the main experiments performed by Ørsted, Ampère, Biot (1774–1862), Savart (1791–1841), Arago (1786–1853), Davy (1778–1829), Wollaston (1766–1828), Berzelius (1779–1848), Berthollet (1748–1822), Schweigger (1779–1857) etc. He then initiated his own investigations. In 1821 and 1822 he published a paper in three parts describing a historical sketch of electromagnetism.[19] He was sceptical in relation to the idea that the electric current consisted in the motion of charged particles. When considering an active apparatus composed of a metallic wire connecting the poles of a Voltaic battery, he expressed his scepticism as follows:[20]

> Those who consider electricity as a fluid, or as two fluids, conceive that a current or currents of electricity are passing between the poles of an active apparatus. There are many arguments in favour of the materiality of electricity, and but few against it; but still it is only a supposition; and it will be as well to remember, while pursuing the subject of electro-magnetism, that we have no proof of the materiality of electricity, or of the existence of any current through the wire.

As regards Ørsted's explanation of the deflection of a magnetic needle by a current of electricity, he said the following:[21] "[...] I have very little to say on M. Ørsted's theory, for I must confess I do not quite understand it." Instead

---

19 [Far21a, Far21b, Far22].
20 [Far21a, p. 196].
21 [Far22, p. 107].

Figure 1.4:
Michael Faraday (1791–1867)
Lenard 1930, p. 224.

of explaining the deflection of the magnetic needle by a conflict of *electricity*, as did Ørsted, Faraday preferred the idea of a *magnetic* force.[22]

Faraday understood Ørsted and Ampère's conceptions of a double current. However, Faraday's emphasis was upon the state of the current carrying wire causing the deflection of the magnetic needle. As regards Ampère's clarification of this state, he said the following:[23]

> *Now as it is in this state that the wire is capable of affecting the magnetic needle, it is very important for the exact comprehension of the theory that a clear and precise idea of its state, or of what is assumed to be its state, should be gained, for on it in fact the whole of the theory is founded. Portions of matter in the same state as this wire, may be said to constitute the materials from which M. Ampere forms, theoretically, not only bar magnets, but even the great magnet of the earth; and we may, therefore, be allowed to expect that a very clear description will first be offered of it. This, however, is not the case, and is, I think, very much to be regretted, since it renders the rest of the theory considerably obscure, for though certainly the highly interesting facts discovered by M. Ampere could have been described, and the general laws and arrangements both in conductors and magnets stated with equal force and effect without any reference to the internal state of the wire, but only to the powers which experiment proves it to be endowed with, yet as M. Ampere has chosen always to refer to the currents in the wire, and in fact founds his theory upon their existence, it became necessary that a current should be described.*

In 1833 Faraday defined current as follows:[24]

> *283. By* current, *I mean anything progressive, whether it be a fluid of electricity, or two fluids moving in opposite directions, or merely vibrations, or, speaking still more generally, progressive forces. By* arrangement, *I understand a local adjustment of particles, or fluids, or forces, not progressive. Many other reasons might be urged in support of the view of a* current *rather than an* arrangement, *but I am anxious to avoid stating unnecessarily what will occur to others at the moment.*

---

22 [Wie88] and [Wie91].
23 [Far22, p. 112].
24 [Far65a, article 283].

Ampère's explanation of Øersted's experiment was based upon a direct interaction between electric currents. This meant an action at a distance, like the Newtonian interaction between two masses or the Coulombian interaction between two electric fluids. Faraday, on the other hand, concentrated his attention on the intervening medium between the interacting bodies. As regards electricity, in particular, he focused his mind on the basic phenomena of *electrostatic* induction, in which we always produce equal and opposite amounts of electricity. According to him, this electrical polarization also happens in the intervening medium between two charged bodies. In 1837 he expressed his views as follows:[25]

> *1162. Amongst the actions of different kinds into which electricity has conventionally been subdivided, there is, I think, none which exceeds, or even equals in importance that called* Induction. *[...] and as the whole effect in the electrolyte appeared to be an action of the particles thrown into a peculiar or polarized state, I was led to suspect that common induction itself was in all cases an action of contiguous particles,*[26] *and that electrical action at a distance (i. e., ordinary inductive action) never occurred except through the influence of the intervening matter. [...] At present I believe ordinary induction in all cases to be an action of contiguous particles consisting in a species of polarity, instead of being an action of either particles or masses at sensible distances; and if this be true, the distinction and establishment of such a truth must be of the greatest consequence to our further progress in the investigation of the nature of electric forces.*

In 1838 Faraday considered electrostatic induction in relation to insulation and conduction. Here he summarized his mature views about the nature of an electric current. Instead of a flow of charged particles, he considered an electric current as the electric polarization of the particles of the medium being communicated sequentially to neighbouring particles. According to his points of view, the particles of a good conductor could not be permanently polarized. He expressed his views as follows:[27]

---

25 [Far65a, articles 1162, 1164 and 1165].
26 [Note by Faraday:] The word *contiguous* is perhaps not the best that might have been used here and elsewhere – for as particles do not touch each other it is not strictly correct. I was induced to employ it, because in its common acceptation it enabled me to state the theory plainly and with facility. By contiguous particles I mean those which are next. – Dec. 1838.
27 [Far65a, articles 1320 and 1338].

1320. Though assumed to be essentially different, yet neither Cavendish nor Poisson attempt to explain by, or even state in, their theories, what the essential difference between insulation and conduction is. Nor have I anything, perhaps, to offer in this respect, except that, according to my view of induction, insulation and conduction depend upon the same molecular action of the dielectrics concerned; are only extreme degrees of one common condition or effect; and in any sufficient mathematical theory of electricity must be taken as cases of the same kind. [...]

1338. To sum up, in some degree, what has been said, I look upon the first effect of an excited body upon neighbouring matters to be the production of a polarized state of their particles, which constitutes induction; and this arises from its action upon the particles in immediate contact with it, which again act upon those contiguous to them, and thus the forces are transferred to a distance. If the induction remain undiminished, then perfect insulation is the consequence; and the higher the polarized condition which the particles can acquire or maintain, the higher is the intensity which may be given to the acting forces. If, on the contrary, the contiguous particles, upon acquiring the polarized state, have the power to communicate their forces, then conduction occurs, and the tension is lowered, conduction being a distinct act of discharge between neighbouring particles. The lower the state of tension at which this discharge between the particles of a body takes place, the better conductor is that body. In this view, insulators may be said to be bodies whose particles can retain the polarized state; whilst conductors are those whose particles cannot be permanently polarized. [...]

1.3.4 The Electric Current According to Maxwell

Maxwell (cf. Figure 2.3, S. 107) followed Faraday's ideas and tried to implement them mathematically. The existence of discontinuous amounts of electricity having a granular structure or an atomistic property did not fit into Maxwell's conception of electricity.[28] Like Faraday, Maxwell rejected the material conception of electricity and denied the existence of electric atoms.[29] We can see this in his discussion of electrolytic conduction. The fundamental law

---

28 [Wie67, p. 109].
29 [Wie08, p. 153].

of electrolysis had been discovered by Faraday. Maxwell presented it with the following words:[30]

> *The number of electrochemical equivalents of an electrolyte which are decomposed by the passage of an electric current during a given time is equal to the number of units of electricity which are transferred by the current in the same time.*

He presented a possible reasonable explanation of this fact in the same article:

> *It is therefore extremely natural to suppose that the currents of the ions are convection currents of electricity, and, in particular, that every molecule of the cation is charged with a certain fixed quantity of positive electricity, which is the same for the molecules of all cations, and that every molecule of the anions is charged with an equal quantity of negative electricity.*

Maxwell presented this atomistic picture of electrolysis as extremely useful in order to understand this phenomenon. But soon after this quotation, he made the following comment revealing his sceptical attitude:[31] *"The electrification of a molecule, however, though easily spoken of, is not so easily conceived."* The same sceptical attitude appears at the end of this article:[32]

> *This theory of molecular charges may serve as a method by which we may remember a good many facts about electrolysis. It is extremely improbable however that when we come to understand the true nature of electrolysis we shall retain in any form the theory of molecular charges, for then we shall have obtained a secure basis on which to form a true theory of electric currents, and so become independent of these provisional theories.*

He discussed the nature of the electric current in many sections of his book *A Treatise of Electricity and Magnetism* of 1873. In article 552 of this work he considered if the energy associated with a current was internal or external to the current:[33]

> *552. It appears, therefore, that a system containing an electric current is a seat of energy of some kind; and since we can form no*

---
30 [Max54a, article 255] and [Max83].
31 [Max54a, article 260, p. 380], [Max83] and [Wie60, p. 154–155].
32 [Max54a, article 260, p. 381], [Max83] and [Wie60, p. 154–155].
33 [Max54b, article 552] and [Max83].

conception of an electric current except as a kinetic phenomenon,[34] *its energy must be kinetic energy, that is to say, the energy which a moving body has in virtue of its motion.*

We have already shewn that the electricity in the wire cannot be considered as the moving body in which we are to find this energy, for the energy of a moving body does not depend on anything external to itself, whereas the presence of other bodies near the current alters its energy.

We are therefore led to enquire whether there may not be some motion going on in the space outside the wire, which is not occupied by the electric current, but in which the electromagnetic effects of the current are manifested.

In Chapter VI of the second volume of his book, articles 568 to 577, Maxwell showed three effects which should manifest themselves if the electric current were composed of inertial masses in motion:[35]

If any action of this kind were discovered, we should be able to regard one of the so-called kinds of electricity, either the positive or the negative kind, as a real substance, and we should be able to describe the electric current as a true motion of this substance in a particular direction. [...]

It appears to me, however, that while we derive great advantage from the recognition of the many analogies between the electric current and a current of material fluid, we must carefully avoid making any assumption not warranted by experimental evidence, and that there is, as yet, no experimental evidence to shew whether the electric current is really a current of a material substance, or a double current, or whether its velocity is great or small as measured in feet per second.

A knowledge of these things would amount to at least the beginnings of a complete dynamical theory of electricity, in which we should regard electrical action, not, as in this treatise, general laws of dynamics, but as the result of known motions of known portions of matter, in which not only the total effects and final results, but the whole intermediate mechanism and details of the motion, are taken as the objects of study.

---

34 [Note by Maxwell:] Faraday, *Exp. Res.* 283.
35 [Max54b, article 574, p. 218] and [Max83].

The experiments performed up to Maxwell's time could not find any of these three effects. Only in the 1910's and 1920's could Tolman and others show the real existence of all these effects, proving in this way the real existence of the inertial mass of the charged carriers, electrons, in metallic conductors.[36] As regards the discovery of the electrons, see also the work by Wiederkehr.[37]

As we can see from this last quotation, in his *Treatise* Maxwell did not regard the electrical current as the result of a flow of charged matter, but only according to the general laws of dynamics.

1.3.5 Weber's Initial Conception of an Electric Current

Contrary to Faraday and Maxwell, Weber accepted the conception of a current as being due to the flow of electricity. Moreover, he considered the current to have a corpuscular or atomic structure.[38] Initially he focused his attention only upon the positive and negative charges of these corpuscles and upon their velocities relative to the wire. But later on he even attributed explicitly inertial masses to these charged particles. In this Section we consider his initial conception of an electric current.

Weber's first major Memoir was published in 1846. It was announced in a paper by Fechner (cf. Figure 2.4, S. 109) of 1845 indicating a connection between the electrodynamic phenomena of Ampère and those of induced currents discovered by Faraday in 1831. Fechner decomposed each current element into two particles with equal and opposite electric charges, moving relative to the wire with velocities of the same magnitude but opposite directions.[39] Weber also accepted initially this supposition of a symmetrical double current. He considered Ampère's force between two current elements. He then supposed each current element as composed of positive and negative charges moving relative to the wire with equal and opposite velocities. This can be seen from his text, namely:[40]

> *If we now direct our attention to the electrical fluids in the two current elements themselves, we have in them like amounts of positive and negative electricity, which, in each element, are in motion in an opposing fashion. This simultaneous opposite motion of positive and negative electricity, as we are accustomed to assume it in all parts of a linear conducting wire, admittedly can not exist in reality,*

---

36 [TOS14], [TKG23], [TMS26], [TS16] and [O'R65].
37 [Wie99].
38 [O'R65, Vol. I, Chapter VII, Section 1: Atomism in electricity].
39 [Fec45].
40 [Web46, Weber's *Werke*, Vol. 3, p. 135–136] and [weba, p. 83].

> yet can be viewed for our purposes as an ideal *motion, which, in the cases we are considering, where it is simply a matter of actions at a distance, represents the actually occurring motions in relation to all the actions to be taken into account, and thereby has the advantage, of subjecting itself better to calculation. The actually occurring lateral motion through which the particles encountering each other in the conducting wire (which latter forms no mathematical line) avoid each other, must be considered as without influence on the actions at a distance, hence it seems permissible for our purpose, to adhere to the foregoing simple view of the matter.*

He expressed this ideal model mathematically in the following way:[41]

> *If we denote by e and e' the positive electrical masses in both elements, and by u and u' their absolute velocities, which have a positive or negative value according to the direction of the current, then $-e$ and $-e'$ will be the negative masses, and $-u$ and $-u'$ their absolute velocities.*

That is, one current element is composed of opposite charges e and $-e$ moving relative to the wire with equal and opposite velocities u and $-u$, respectively. In the same work he expressed this fact with the following words:[42] *"[...] each current element should contain the same amount of positive and negative electricity, and both should flow through the element with the same velocity, but in opposite directions."*

Weber also presented this idea of a double current in his works of 1855–1857 with his friend Rudolph Kohlrausch (1809–1858) on the measurement of Weber's constant.[43] Here we quote from the work of 1856, our words in square brackets:[44]

> *We imagine that in the bodies constituting the [galvanic] circuit, their neutral electricity is in motion, in the manner that their entire positive component pushes around in the one direction in closed, continuous circles, the negative in the opposite direction. [...] This measure, which will be called the mechanical measure of current intensity, thus sets as the unit, the intensity of those currents which*

---

41 [Web46, Weber's *Werke*, Vol. 3, p. 139] and [weba, p. 85].
42 [Web46, Weber's *Werke*, Vol. 3, p. 203] and [weba, p. 133].
43 [Web55, p. 594 of Weber's *Werke*], [WK56, p. 597–598 of Weber's *Werke*] with English translation in [WK03, p. 287-288], and [KW57, p. 614–115, 619–620 and 648].
44 [WK56, p. 597-598 of Weber's *Werke*] with English translation in [WK03, p. 287–288].

> *arise when, in the unit of time, the unit of free positive electricity flows in the one direction, an equal amount of negative electricity in the opposite direction, through that cross-section of the circuit.*

In his work of 1846 Weber expressed his belief that this supposed uniform motion was only an idealization. As a matter of fact, even for steady currents, the real motions of the charges were not constant:[45]

> *In the method for determining galvanic current given in Section 19, on which the law describing two electrical masses acting on one another at a distance is based, instead of the actual current, in which the velocity of the flowing electricity probably fluctuates in its passage from one ponderable particle to the other in a steady alternation, an ideal current of uniform velocity is assumed. This substitution was necessary to simplify the treatment, and it seems permissible because it is simply a question of an action at a distance.*

The expression which has been translated as "from one ponderable particle to the other," appears in German[46] as "von einem ponderablen Theilchen zum anderen." In his paper of 1846 Weber utilized many times the adjective *ponderable*: for bodies, for wire elements or current carriers, for conductors, for particles and for molecules. He probably did mean with this adjective "having appreciable weight," as distinct from electrical particles belonging to the current carrying elements which would have virtually no mass, or whose mass would be negligible. For these last particles he utilized the adjective *imponderable*:[47]

> *Ampère's law leaves nothing to be desired, when it deals with the reciprocal actions of conducting wires, whose currents posses a constant intensity, and which are fixed in their positions with respect to one another; as soon as changes in the intensity of the current take place, however, or the conducting wires are moved with respect to one another, Ampère's law gives no complete and sufficient account; namely, in that case, it merely makes known the actions which take place on the ponderable wire element, but not the actions which take place on the imponderable electricity contained therein. Therefore, from this it follows, that this law holds only as a particular law, and can be only provisionally taken as a fundamental law; it still requires a definitive law with truly general validity, applicable to all electrodynamic phenomena, to replace it.*

---
45 [Web46, Weber's *Werke*, Vol. 3, p. 207–208] and [weba, p. 137].
46 [Web46, Weber's *Werke*, Vol. 3, p. 207–208].
47 [Web46, Weber's *Werke*, Vol. 3, p. 133] and [weba, p. 82].

It is amazing to note that even in this first article Weber already departed from the models of Ørsted and Ampère by considering the possibility of positive and negative charges having different masses, as will be shown in the next quotation. If this were the case, they would move inside a conductor carrying a constant current with velocities having different magnitudes:[48]

> *By a galvanic current, as opposed to other electrical motions not comprised under this name, should be understood a motion of the electricity in a closed conductor, such that the same amounts of positive and negative electricity flow through all its cross-sections simultaneously in the opposite directions. This equality of the positive and negative electricity flowing through does not necessarily presuppose the equality of the moving positive and negative masses, which was previously assumed, but rather, it can exist even when the latter are of unequal magnitudes, if the larger mass flows slower, the smaller one faster.*

This is a great insight which he developed later on. Before going to his mature conception of an electric current, we must consider the nature of the molecular currents. Weber's conception of these molecular currents represented his first step towards a planetary model of the atom.

1.4 The Nature of the Molecular Currents

1.4.1 The Molecular Currents According to Ampère

It was known for centuries that the Earth oriented a magnetic needle along the local magnetic meridian. If the magnetic needle were released at rest oriented outside the local plane of the magnetic meridian, it would be deflected by the Earth towards the local magnetic meridian. Ørsted's experiment of 1820 showed the deflection of a magnetic needle by a nearby electric current. Ampère followed this idea a step further and supposed that the magnetic properties of the Earth might be due to electric currents circulating inside the Earth in planes parallel to the equator.[49] He expressed his reasoning as follows, as quoted in Darrigol:[50]

---

48 [Web46, Weber's *Werke*, Vol. 3, p. 204] and [weba, p. 134].
49 [Wie88], [Ste03] and [Ste05].
50 [Dar00, p. 6]. See also [Amp20b, p. 202–203], [Amp65a, p. 152] and [Blo82, p. 72–73].

Granted that the order in which two facts have been discovered does not make any difference in the available analogies, we could suppose that before we knew about the South-North orientation of a magnetized needle, we already knew the needle's property of taking a perpendicular position to an electric current [...]. Then, for one who tries to explain the South-North orientation, would not it be the simplest idea to assume in the Earth an electric current?

As a bar magnet also deflects a magnetic needle, the next step for Ampère was to suppose the existence of electric currents encircling the magnetic axis of the magnetic bar, within planes perpendicular to that axis. He then postulated that the magnetic interactions between two magnets, or between a magnetic needle and the Earth, were due only to interactions between the electric currents existing inside these bodies. This suggested a test of this hypothesis, namely, to look for a direct interaction between two electric currents. Until this time no interaction of this type had ever been observed.

In 1820 Ampère coiled two spirals and placed them in parallel planes facing one another. One of these spirals was fixed in the laboratory, while the other could approach or move away from the first one. By passing an electric current through both spirals, he was able to show an attraction or repulsion between them depending upon the directions of the currents.[51] This crucial experiment created a new field of research, that of interaction between two current carrying wires. Following this experiment, he also showed that two parallel wires attract one another when the currents flow in the same direction, repelling one another when they flow in opposite directions.[52]

In order to characterize this new field of research, Ampère coined[53] the names "electrostatics" and "electrodynamics". The field of electrostatics described the interaction between charges at rest, while that of electrodynamics characterized the interaction between charges in motion, like the interaction between two current carrying wires.

Ampère's speculations of 1820 related to macroscopic electric currents inside magnets were criticized by his friend Augustin Fresnel (1788–1827).[54] Fresnel notes were only published in 1885.[55] Fresnel, for instance, pointed out that if the magnetism of a bar magnet was due to electric circuits of macroscopic dimensions, thermal effects should be detectable within the magnet. But

---

51 [Amp20b].
52 [Amp20a].
53 [Amp22a] and [Blo82, p. 78].
54 [Blo82, p. 98–101 and 118–125], [Hof82, p. 334 and 343], [Hof87] and [Hof96, p. 282–290].
55 [Fre85a] and [Fre85b].

these effects were not observed. In order to avoid Fresnel's objection, Ampère adopted his suggestion of January of 1821 that the currents responsible for the magnetic phenomena in magnetized bars were restricted to molecular dimensions around the axes of each molecule or particle composing the bar.

In his main work of 1826, Ampère expressed this conception of molecular currents with the following words:[56]

> *In order to justify the manner in which I have conceived magnetic phenomena, regarding magnets as assemblies of electric currents forming minute circuits round their particles, [...] and the manner in which I have explained magnetic phenomena associated with electric currents forming very small closed circuits round particles of a magnetized body, [...]*

1.4.2 On the Origins of the Resistance of Conductors According to Weber

In 1852 Weber published a second major Memoir, namely, *Electrodynamic Measurements Relating Specially to Resistance Measurements*.[57] A constant current arises in a conductor when an electromotive force or tension is applied upon it, according to Ohm's law of 1826.[58] Weber wanted to understand the origin of this resistance in terms of microscopic forces acting upon the conducting charges. According to Weber, this resistance in a metallic conductor arises due to a continuous connection and separation of positive and negative charges encountering one another in a double current. This is somewhat similar to Ørsted's conflict or to Ampère's model of an electric current.

Inside conductors acted upon by electromotive forces, the mobile charges experience a resistance to their progressive motion along the direction of the wire. In order to have a constant current inside a resistive wire, it is necessary a constant electromotive force, or a constant tension between the extremities of the conductor. As an atomist, Weber was not satisfied with the mere statement of this empirical law. He wanted to find or to clarify the essence of the resistance in terms of atomic or molecular processes inside the wire. In his work of 1852 Weber wanted to find the origins of the resistance of metals. He expressed himself as follows, our words in square brackets:[59]

---

56 [Amp26, p. 105 and 116], [Amp23, p. 277 and 288] and [Amp65b, p. 192 and 196].
57 [Web52b].
58 [Ohm26], [fec] and [ram].
59 [Web52b, Section 36, p. 400–403]:
    36. Ueber die Ursachen des Widerstands der Leiter.

Zu einer vollständigen Kenntniss des Widerstands genügt es nicht, die Grösse des Widerstands aus seinen Wirkungen zu definiren, d. i. aus der Stärke des durch eine gegebene elektromotorische Kraft hervorgebrachten Stroms, sondern es gehört auch dazu, die Grösse des Widerstands aus ihren Ursachen zu definiren. Ohne diese wesentliche Ergänzung ist unsere Kenntniss von dem Wesen des Widerstands mangelhaft, und die ermittelte Grösse desselben ist eine blosse Hülfsgrösse der Elektrodynamik, deren wahre physische Bedeutung noch unbekannt ist. Wenn nun der Widerstand bisher blos nach seinen Wirkungen betrachtet worden ist, so liegt der Grund davon darin, dass über die Ursachen desselben bisher noch gar nichts Wesentliches ermittelt worden ist. Es ist blos die Abhängigkeit des Widerstands von den äusseren Dimensionen des Leiters, nämlich von seiner Länge und von seinem Querschnitt, ermittelt worden, aber diese Abhängigkeit betrifft blos den absoluten Widerstand eines Leitungsdrahts und hat keine Beziehung auf den specifischen Widerstand des leitenden Metalls, über dessen Ursachen gar nichts bekannt ist. Diese Ursachen scheinen so tief in der Natur der Körper verborgen zu liegen, dass sie auf den bisherigen Wegen der Forschung unzugänglich sind. Kurz, die Frage nach den Ursachen des galvanischen Widerstands führt zu einem noch ganz unangebauten Gebiete der Wissenschaft. Ich werde mich daher nur auf eine einzelne Erörterung beschränken, nämlich darüber, in welcher Beziehung dieser Widerstand mit der Natur der elektrischen Fluida selbst, wie dieselben definirt worden sind, und mit deren Verhalten im elektrischen Doppelstrome stehe, wie dasselbe nach der gewöhnlichen Vorstellung auch hier immer angenommen und festgehalten worden ist.

Die Frage nach den Ursachen des Widerstands lässt sich zunächst specieller darauf richten, in wie weit diese Ursachen in dem ponderablen Träger des Stroms, und in wie weit dieselben in den darin enthaltenen elektrischen Fluidis liegen. Dass die Gegenwart der ponderablen Theile die Kanäle, durch welche die elektrischen Fluida strömen, mehr oder weniger beengen, und dadurch auf die elektrische Strömung Einfluss haben können, leuchtet von selbst ein; es fragt sich aber, ob diese Ursache zur Erklärung des Widerstands allein schon genüge. Diese Ursache des Widerstands würde blos die Masse des elektrischen Fluidums beschränken, welche an der Strömung Theil nehmen könnte. Es liegt aber in dem Wesen des Widerstands, wie wir ihn aus seinen Wirkungen kennen, dass durch die Grösse des Widerstands nicht blos die Masse des elektrischen Fluidums beschränkt wird, welche an der Strombewegung Theil nimmt, sondern, dass auch die Bewegung selbst beschränkt wird. Diese Beschränkung der Bewegung selbst kann aber ihren Grund in der blossen Gegenwart der ponderablen Theile nicht haben, sondern setzt nothwendig Kräfte voraus, welche den fortwirkenden elektromotorischen Kräften der Kette das Gleichgewicht halten, weil ohnedem jene Kräfte die elektrischen Fluida in ihrer Bewegung immerfort beschleunigen müssten, was bei einem gleichförmigen und beharrlichen Strome nicht der Fall ist.

Es fragt sich also ferner, woher die Kräfte rühren, welche bei einem gleichförmigen und beharrlichen Strome den fortwirkenden elektromotorischen Kräften das Gleichgewicht halten und dadurch eine fernere Beschleunigung der elektrischen Fluida in ihrer Bewegung verhindern? Sind diese Kräfte rein elektrische Kräfte, oder sind es Kräfte, welche die ponderablen Theile auf die elektrischen Fluida, die an ihnen vorbeigehen, ausüben? Setzen wir in dem galvanischen Strome, wie wir es stets gethan haben, zwei elektrische Fluida voraus, die gleichzeitig durch denselben Leiter in entgegensetzten Richtungen strömen, so liegt es sehr nahe, eine Ursache des Widerstands für die Bewegung jedes Fluidums in dem ihm entgegenkommenden Fluidum zu suchen. Das positive und das negative Fluidum werden nämlich in dem Augenblicke der Begegnung sich zu neutralem Gemische verbinden, und so leicht auch diese neutrale Verbindung wieder zu scheiden sein möge, so wird doch eine solche neue Scheidung nur durch eine neue elektromotorische Kraft erfolgen

36. *On the Origins of the Resistance of Conductors*

For a complete knowledge about the resistance [of conductors] it is not enough to define the amount of resistance from its effects, that is, from the amount of the produced current through a given electromotive force, as it is also necessary to define the resistance from its origins. Without this basic extension, our knowledge about the essence of the resistance is incomplete, and the determined amount of resistance is a bare auxiliary quantity of electrodynamics, whose real physical meaning is still unknown. The reason why the essence of its origins has not yet been determined, comes from the fact that up to now the resistance has been considered only according to its effects. It is determined merely the dependence of the resistance upon the external dimensions of the conductor, that is, its length and cross-section, but this dependency concerns only the absolute

---

können, und nicht in Folge einer Beharrung derjenigen Bewegungen, welche beide Fluida vor ihrer Vereinigung besassen, weil diese durch ihre Begegnung und Verbindung mit einander als aufgehoben betrachtet werden muss. Es geht daraus hervor, dass, während jedem Fluidum für sich bei seinen Bewegungen Beharrung zugeschrieben werden muss, beiden Fluidis zusammen bei ihrer Bewegung im Doppelstrome keine Beharrung zukommt. Wenn aber auch dieser Grund, warum den elektrischen Fluidis bei ihrer Bewegung im Doppelstrome keine Beharrung zukommt, der richtige ist, so gewinnt man doch dadurch noch keine deutliche Einsicht in den Hergang selbst, so lange die Kräfte unbekannt sind, welche die Verbindung und Vereinigung der elektrischen Fluida bei ihrer Begegnung bewirken, und welche bei ihrer wiederholten Scheidung überwunden werden müssen. Es fragt sich, ob dabei noch andere Kräfte in Betracht kommen, als diejenigen, welche durch das allgemeine elektrische Grundgesetz schon bestimmt sind, z. B. ob dabei besondere Molekularkräfte der elektrischen Fluida wirksam sind. Wäre dies nicht der Fall, so müsste der Hergang bei der abwechselnden Verbindung und Scheidung der elektrischen Fluida im Doppelstrome nach dem bekannten Grundgesetze der elektrischen Wirkung genauer bestimmt werden. Ohne eine solche genauere Bestimmung lässt sich im Allgemeinen nur mit einiger Wahrscheinlichkeit annehmen, dass die Intensität eines elektrischen Doppelstroms ausser von der Masse der elektrischen Fluida, welche an der Strömung Theil nimmt, von der Zahl der Scheidungen abhänge, welche in bestimmter Zeit erfolgen, und dass die Zahl dieser Scheindungen der während dieser Zeit fortwirkenden elektromotorischen Kraft proportional sein müsse. Ergäbe sich z. B., dass durch gleiche elektromotorische Kraft jedes elektrische Theilchen in gleicher Zeit immer eine gleiche Zahl Verbindungen und Scheidungen erlitte und dadurch eine gleiche Wegstrecke fortgeführt würde, so wäre die Stromgeschwindigkeit u für gleiche elektromotorische Kraft immer die nämliche, und es würde dann die Stromintensität für gleiche elektromotorische Kraft blos mit der Menge der Elektricität e variiren, welche auf einer solchen Wegstrecke (z. B. in der Längeneinheit des Leiters) enthalten wäre, und zwar proportional damit sein, woraus hervorginge, dass der sogenannte Widerstand gleichfalls nur mit e variirte und zwar dem Werthe von e umgekehrt proportional wäre, welches derjenige Fall ist, welcher am Ende des vorigen Artikels als Erläuterung angeführt wurde.

resistance of a conducting wire and has no relation with the specific resistance of the conducting metal, so that about its origins nothing is known. These origins seem to be so deeply concealed in the nature of the bodies, that they seem to be impenetrable from the route of research pursued up to now. In short, the question about the origins of the galvanic resistances leads to a region of science which has not yet been developed. I will restrict myself to a single argument, that is, what is the relation between this resistance and the nature of the electric fluids, in the way these were defined, and with their behavior in a double current, in the way this [current] according to the usual conception will also here be assumed and adhered to.

The question about the origins of the resistance can be at first specifically addressed in trying to know how much of it is due to the ponderable carrier of the current and how much of it is due to the electric fluids contained in it. It is self evident that the presence of canals in the ponderable parts, which are more or less narrow and through which flow the electric fluids, can have an influence upon the electric current; however, it can be asked if only this is enough to clarify the origin of the resistance. This origin of the resistance would only limit the quantity of the electric fluid, which could take part in the current. However, it belongs to the essence of the resistance, as we know from its effects, that not only the quantity of the electric fluid, which take part in the current motion, is limited by the amount of resistance, but also that the motion itself is constrained. This limitation of the proper motion can not have its origin in the mere presence of the ponderable parts, but presupposes some necessary forces, which maintain the equilibrium of the continuously acting electromotive forces of the circuit, because without these forces the electric fluids would always have an accelerated motion, which is not the case by an uniform and constant current.

Moreover, it can also be asked in what consist these forces which, in an uniform and constant current, maintain the equilibrium of the continuously acting electromotive forces and so prevent the acceleration of the electric fluid in its motion. Are these forces purely electric forces, or are they forces which the ponderable particles exert upon the electric fluid which pass by them? When we presuppose in a galvanic current, as we have always made, two electric fluids which flow simultaneously through the same conductor in opposite directions, then it suggests itself to search for an origin of the re-

sistance in the motion of the [positive] fluid meeting an oncoming [negative] fluid. That is to say, the positive and negative fluids will combine in neutral mixtures at the moment of the encounter. This neutral connection may also come once more to a new separation. This means that a new decomposition will only be possible through a new electromotive force, and not as a consequence of the maintenance of those motions which both fluids had before their encounter, because [these motions] must be considered as abolished through their encounter and connection. It emerges from this that, while it can be attributed to each fluid alone a maintenance of its motion, it does not come any steady condition for both fluids moving together in a double current. However, when this is the correct origin why the electric fluids by their motion in a double current do not give rise to any steady condition, even so we gain no clear comprehension in the proper course of events, so long as the forces which cause the connection and aggregation of the electric fluids remain unknown, and which must be overcome during a new decomposition. It can be asked if we need to take into account other forces, beyond those which were already determined though the general fundamental electric law, for instance, if special molecular forces of the electric fluids are active. If this were not the case, then the course of events during the repeated composition and decomposition of the electric fluids in a double current would be precisely determined by the known fundamental law of electric action. Without this precise determination we remain in general only with some assumed presumptions, that the intensity of a double electric current depends, beyond the quantity of the electric fluids which take part into the current, upon the number of decompositions which happen in a certain time, and that the number of these decompositions must be proportional to the acting electromotive force. For instance, if it arose through equal electromotive force, that each electric particle at the same time always made an equal amount of compositions and decompositions and so an equal distance were pursued, then for equal electromotive force the current velocity u would always be the same, and then the current intensity for the same electromotive force would vary only with the amount of electricity e, which for such a distance (for instance, in a unit length of the conductor) is contained in it and, indeed, would be proportional to it, from which it would come, that the so called resistance would only vary as well with e and, indeed, *it would be inversely proportional with e, which is that case, which was introduced as a clarification at the final of the previous Section.*

That is, according to Weber's conception, the origin of the resistance of metals is related with the encounter of positive and negative charges moving in opposite directions inside a current carrying conductor.

1.4.3 The Molecular Currents According to Weber

In 1839 and 1841 Weber began some experiments related to unipolar induction, a name which he coined for an effect which had first been observed by Faraday in 1831. Until this time, Weber had considered all magnetic effects as due to the existence of magnetic dipoles (that is, a dipole consisting of a north and a south magnetic fluids separated by a small distance). In his works of 1839 and 1841, he began to consider the alternative view of Ampère's electrodynamic theory of magnetic phenomena. Accordingly, instead of a magnetic dipole, Weber considered the possible existence of constant galvanic currents inside magnets. According to his first works of 1839 and 1841, Weber considered that the phenomenon of unipolar induction could only be explained supposing the existence of magnetic fluids, but not with Ampère's hypothesis about the existence of constant galvanic currents inside magnets.[60] In 1852 Weber corrected himself, showing that also Ampère's hypothesis could explain unipolar induction phenomena.[61] In any event, at least around 1839 Weber began to think seriously about the existence of Ampère's molecular currents. This conception of molecular currents had a great influence upon his future work.

In 1848 and 1852 Weber showed that Faraday's discovery of diamagnetic phenomena (1845) could only be explained supposing the existence of Ampère's molecular currents, but not supposing the existence of magnetic fluids.[62]

In Section 1.4.2 it was shown that according to Weber's conception of 1852, the origin of the resistance of metals was related with the encounter of positive and negative charges moving in opposite directions inside a current carrying conductor. If this were the case, then it would raise a problem with Ampère's molecular currents. According to Ohm's law, if the electromotive force disappears, the electric current also goes to zero. How would it be possible to have a *constant or permanent* molecular current, as suggested by Ampère, if they were composed of positive and negative charges moving in opposite

---

60 [Web39, Weber's *Werke*, p. 171–175] and [Web41].
61 [Web52a, Weber's *Werke*, p. 535–538].
62 [Web48b, Weber's *Werke*, p. 264–268] (English translation in [Web66b]), [Web52a, Weber's *Werke*, p. 535–538] and [Web52c, Weber's *Werke*, p. 568–570] (English translation in [Web66a]).

directions in circular orbits around the molecule and, therefore, meeting one another? If this were the case, these molecular currents would also be resistive. Therefore, it would be necessary a microscopic source of electromotive force in order to keep these molecular currents with a constant intensity. Weber did not want to introduce this microscopic source of electromotive force. On the other hand, he wanted to maintain Ampère's idea of a molecular current because he needed this concept in his theories of magnetism and diamagnetism. The combination of Ohm's law with Ampère's molecular current created a problem for Weber. But in 1852 he found a fascinating solution to solve this problem. We present here his relevant line of reasoning:[63]

> *If the origin of the resistance were really contained in the alternate composition and decomposition of the electric fluids during their encounter in a double current, then this would imply the impossibility of a constant double current without the existence of a continuous external electromotive force, and then we could ask, how would this be reconcilable with the supposition of constant molecular currents for the explanation of magnetic and diamagnetic phenomena. The possibility of these molecular currents must then necessarily depend upon an action of the ponderable molecule, through which the orbits of the electric fluids moving around each molecule in opposite directions were maintained separate from one another, in which, for example, a fluid would describe a narrower circular orbit around the molecule while the other fluid would describe a larger circular orbit, in such a way that both fluids could never encounter and unite with one another during their motions.*

That is, if the charged particles moved in orbits having different radii, they would never meet one another. Therefore, they could continue their orbits

---

63 [Web52b, Weber's *Werke*, p. 403]:
    Sollte in der abwechselnden Verbindung und Scheidung der elektrischen Fluida bei ihrer Begegnung im Doppelstrome die Ursache des Widerstands wirklich enthalten sein, so würde daraus ferner die Unmöglichkeit eines *beharrlichen* Doppelstroms ohne fortwirkende äussere elektromotorische Kraft folgen, und es würde sich dann fragen, wie damit die Annahme von *beharrlichen Molekularströmen* zu Erklärung der magnetischen und diamagnetischen Erscheinungen verträglich wäre. Die Möglichkeit solcher Molekularströme müsste dann nothwendig auf einer Wirkung der ponderablen Moleküle beruhen, durch welche die Bahnen der in entgegengesetzten Richtungen um jene Moleküle sich bewegenden elektrischen Fluida von einander getrennt erhalten würden, indem z. B. das eine Fluidum eine engere Kreisbahn, das andere Fluidum eine weitere Kreisbahn um das Moleküle beschriebe, sodass die beiden Fluida sich bei ihren Bewegungen nirgends begegnen und vereinigen könnten.

without the existence of an applied electromotive force. After all, no resistance would arise due to their motions in opposite directions, as they would never meet one another! By following this idea a step further, he now considered that the positive charges remained fixed with the ponderable atoms of the metal. In a first model these atoms were considered as uniformly distributed along a straight line. Around each of these positive atoms, the negative charges orbited in circular or elliptical orbits, like planets around the Sun:[64]

> *In order to clarify the order of events which happen by the alternate composition and decomposition of the electric fluids in a double current, as would be conveyed by the fundamental law of electrical action without consideration of special molecular forces, the following consideration is useful. Consider that in A, B, C, ..., there are positive electrical masses, which will be assumed initially to remain fixed in the places where they are found. In a there is actually a negative and mobile electrical mass, upon which the nearby positive mass in A acts so strongly, that the actions of the farther masses at B, C, ..., can be considered negligible. The masses at A and at a act upon one another with a force which depends upon their magnitudes, distance, relative velocity and its variation; however, here we assume the simplification that the correction in relation to the electrostatic force (which depends upon the masses and their distance) arising from the relative velocity and its variation is so small, that they can also be neglected. It follows from these suppositions that, when no other force acts upon the mass a, this mass must follow the*

---

64 [Web52b, Weber's *Werke*, p. 403–404]:
Zur Erläuterung des Hergangs bei der abwechselnden Verbindung und Scheidung der elektrischen Fluida im Doppelstrome, wie er aus dem Grundgesetze der elektrischen Wirkung ohne Zuziehung besonderer Molekularkräfte dieser Fluida abzuleiten wäre, diene folgende Betrachtung. In A, B, C ... seien positiv elektrische Massen, von denen zunächst angenommen werden möge, dass sie an den Orten, wo sie sich befinden, festgehalten würden. In a befinde sich gegenwärtig eine bewegliche negative elektrische Masse, auf welche die benachbarte positive Masse in A so stark wirke, dass dagegen die Wirkung der entfernten Massen in B, C ... vernachlässigt werden könne. Die Massen in A und a wirken auf einander mit einer Kraft, die von ihrer Grösse, Entfernung, relativen Geschwindigkeit und deren Aenderung abhängt; indess möge hier der Einfachheit wegen angenommen werden, dass die aus der relativen Geschwindigkeit und deren Aenderung sich ergebende Korrektion der elektrostatischen (von den Massen und der Entfernung abhängigen) Kraft gegen diese letztere so gering sei, dass sie ebenfalls vernachlässigt werden dürfte. Unter diesen Voraussetzungen folgt, dass, wenn keine andere Kraft auf die Masse in a wirkt, diese Masse den Gesetzen der Bewegung durch Centralkräfte, welche dem Quadrat der Entfernung verkehrt proportional sind, folgen müsse. Die Masse in a wird folglich nach den Kepler'schen Gesetzen z. B. eine elliptische Bahn um A beschreiben.

laws of motion through central forces, which are inversely proportional to the square of the distance. Therefore, the mass in a will describe for example an elliptical orbit according to Kepler's laws.

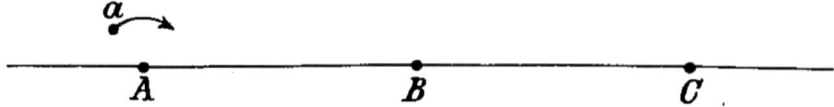

Figure 1.5:
Weber's conception of Ampère's molecular current.
This represents Weber's simplest planetary model of the atom.
In this idealized conception, a negative charged particle a follows
an elliptical orbit around a positive ponderable electrical mass A.
[Web52b, Weber's *Werke*, p. 403].

We can consider this Weberian corpuscular conception of Ampère's molecular current as his simplest planetary model of the atom. It is a transformation or modification of Ampère's original model of a molecular current. Eduard Riecke (1845–1915), Weber's successor at Göttingen University, mentioned that, by following this line of reasoning, Weber disintegrated Ampère's rings, that is, his molecular currents, into a system of electrical satellites.[65] We can even date the specific beginning of Weber's planetary model in this fundamental work of 1852, *Electrodynamic Measurements Relating Specially to Resistance Measurements*.[66]

1.4.4 Further Developments of Weber's Conception about Ampère's Molecular Currents

In 1862 Weber presented once more a corpuscular or planetary model of Ampère's molecular current. The only difference as regards his work of 10 years earlier, is that now he reversed the signs of the stationary and mobile electrical charges. That is, now he supposed a positive charge orbiting around a ponderable negative molecule, although he was open to both possibilities. During

---

65 [Rie92, p. 25].
66 [Web52b].

Weber's life time it was not yet possible to discover which model would be more compatible with nature. He presented this model as follows:[67]

> *When we suppose, for instance, the negative fluid as rigidly connected with the molecule, and consider only the positive fluid in molecular current, or vice versa (a conception which recommends itself, as it is consistent with the persistence of the molecular currents without electromotive forces), [...]*

In 1871 he presented once more this planetary model of Ampère's molecular current, emphasizing its advantages.[68] In this work, he considered the motion of two electrical particles (*elektrische Theilchen*) with charges e and e' interacting with one another according to his force law. Weber attributed inertial masses ε and ε' for these charged particles, considering these masses much smaller than the masses of the ponderable atoms or molecules. He had already introduced inertial masses for the electric fluids in 1864, as will be discussed in Section 1.6.2. He also considered the possibility that each of these electrical particles might adhere to a ponderable atom or molecule of mass m. These charged atoms or molecules are called *ions* nowadays. As regards the order of magnitudes of ε and m, he made the following extremely interesting comment:[69]

> *In many cases the electrical mass ε is connected with a ponderable mass m, so that it is impossible for it to be moved independently of it; in such cases, only the combined mass $m + \varepsilon$ comes into account, and in general ε may be regarded as vanishingly small in comparison with m. Consequently it is only seldom that the masses ε, ε' have to be considered.*

It is amazing that even without knowing the order of magnitude of ε, Weber already suspected that it would be much smaller than the mass of the ponderable particles. Along the paper, he considered the motion of two particles of charges e and e' interacting with one another according to his force law. As usual, he attributed inertial masses ε and ε' to these particles. In Section 15

---

67 [Web62, p. 95]:
   Denkt man sich dann also z. B. das negative Fluidum mit dem Molekule als fest verbunden, und das positive Fluidum allein in Molekularströmung begriffen, oder umgekehrt (eine Vorstellungswese, welche sich dadurch empfiehlt, dass sie mit der Beharrung der Molekularströme ohne elektromotorische Kräfte bestehen kann) [...]
68 [Web71, Weber's *Werke*, p. 281–285] and [Web72, p. 132–136].
69 [Web71, p. 251 of Weber's *Werke*] and [Web72, p. 3].

of the paper, Weber gave another name to these electrical particles, namely, *electrical atoms (elektrische Atome)*.[70] He then made a similar comment:

> *For in the general distribution of electricity it must be assumed that an atom of electricity adheres to each ponderable atom. But if atoms of electricity adhere firmly to ponderable atoms, nothing will be altered in the relations of the electrical atoms except the masses which have to be moved by the forces acting on the electrical atoms. But in the preceding developments the masses are left undetermined, and are simply denoted by ε and ε'; while the electrical particles themselves, to which the masses ε and ε' belong, are determined, without a knowledge of the values ε and ε', by the measurable quantities e and e'. If now we take the values of ε and ε' so great as to include the masses of the ponderable atoms adhering to the electrical atoms, all the results that have been arrived at in reference first of all to electrical atoms merely, may also be applied to the ponderable atoms combined with the electrical atoms.*

That is, in this case the mass of this charged ponderable atom (which nowadays would be called an ion, with a total mass $\varepsilon + m$ or $\varepsilon' + m'$) was also represented by ε or by ε'.

He considered the motion of two charges of opposite sign moving relative to one another according to his fundamental force law. He showed that there was one solution for this problem in which these charged particles orbited around one another while keeping a constant distance of separation. He said the following in Section 17 of this work, which was devoted to a discussion of Ampère's molecular currents:[71]

> *The relation between the particles in respect of their participation in the motion depends upon the ratio of their masses ε and ε'; and, according to section 15, the values of ε and ε' must include the masses of the ponderable atoms adhering to the electrical atoms. Let e be the positive electrical particle, and let the negative particle be equal and opposite to it, and let it therefore be denoted by $-e$ (instead of e'). Now let a ponderable atom adhere to the latter only, whereby its mass is so much increased that the mass of the positive particle becomes negligible in comparison. The particle $-e$ may then be regarded as being at rest, and the particle $+e$ alone as being in motion around the particle $-e$.*

---

70 [Web71, p. 279 of Weber's *Werke*] and [Web72, p. 130].
71 [Web71, Weber's *Werke*, p. 281] and [Web72, p. 132].

In the conclusion of this Section, Weber showed that this model was completely compatible with Ampère's molecular currents for two main reasons, namely:[72]

> Hence it appears that an electrical particle $+e$ moving in a circle about the electrical particle $-e$ exerts upon all galvanic currents the same effects as those assumed by Ampère in the case of his molecular currents.
> The molecular currents assumed by Ampère, however, differ essentially from all other galvanic currents in this respect, that, according to Ampère's assumption, they continue *without electromotive force*; whereas all other galvanic currents, in accordance with Ohm's law, are proportional to the electromotive force, and cease when the electromotive force vanishes. But it is evident that the electrical particle $+e$, spoken of above, must of itself, without electromotive force, continue indefinitely its rotatory motion about the particle $-e$, and therefore must correspond entirely with the molecular currents assumed by Ampère in this respect also.
> We accordingly obtain in this way, as a deduction from the laws of the molecular state of aggregation of two dissimilar electrical particles, developed in the preceding section, a simple construction for the molecular currents assumed by Ampère without proof that their existence was possible.

All of these quotations indicate clearly that one of the origins of Weber's planetary model of the atom was related with his conception about the nature of Ampère's molecular currents. His atomistic point of view gave him the key to devise a planetary model in which a charged particle would orbit around a charged particle of opposite sign. As these charges did not encounter one another during their motion, there was no resistance and this orbit would continue indefinitely without the presence of any electromotive force. Moreover, Weber was able to show mathematically that, according to his own force law between point charges, this planetary model produced the same force upon an external galvanic current as the force produced by Ampère molecular current and expressed by Ampère's force law between current elements.

Weber's points of view about Ampère's molecular current were then the main inspiration which led to his mature conception about the nature of an electric current. Weber's planetary model of an Ampèrian molecular current solved the problem of its permanent state without a microscopic source of electromotive

---

72 [Web71, Weber's *Werke*, p. 285] and [Web72, p. 135–136].

force. This planetary conception was based upon the idea of a charge of one sign being fixed with the weighty molecule. At the same time, a charge of the other sign and negligible mass, would orbit around the molecule like a planet around the Sun. The evidence at his disposal at this time did not permit him to decide if this flow was composed of a positive or of a negative charge. For this reason, he let this question open for the time being, being open to both possibilities.

1.5 The Evolution of Weber's Conception of an Electric Current: From a Double Current to a Simple Current

In 1846 Weber had assumed a double current of positive and negative charges moving relative to the wire with equal and opposite velocities, as seen in Section 1.3.5. In 1852 he considered another model for an electric current. He now considered that the positive charges remained fixed with the ponderable atoms of the metal. In a first model these atoms were considered as uniformly distributed along a straight line. The negative charges orbited around each of these positive atoms in circular or elliptical orbits, like planets around the Sun. Let us consider a specific negative charge orbiting around a first positive atom fixed in the lattice of the metal. If a tension or electromotive force was applied along the conductor, this negative charge would begin to move along a spiral line around the first positive atom, with increasing radius. Eventually it would come into the sphere of influence of another positive atom, beginning to orbit around it. If the applied electromotive force continued to be applied for some time, the negative charge would follow once more another spiral line around this second positive atom, until it would come into the sphere of influence of a third positive atom, beginning to orbit around it. This motion would go on continuously, producing on average a uniform motion of the negative charges along the direction of the conductor. According to Weber, the resistance of the metals had its origin in the centripetal forces exerted by the positive charges. If the electromotive force did stop, the negative charges would remain orbiting the last positive atom around which they were circling previously. His description was presented as follows:[73]

---

[73] [Web52b, p. 404–405]:

Es wird aber eine Störung in dieser Bewegung der betrachteten Masse um A eintreten, sobald ausser der Centralkräft eine elektromotorische Kraft parallel mit der Linie AB mit konstanter Intensität auf die betrachtete Masse wirkt. Die Elemente der bisherigen ellip-

However, there will arise a perturbation in this motion of the considered mass which moves around A [see Figure 1.5], as soon as an electromotive force parallel with the line AB acts upon this mass, beyond the central force. The elements of the previous elliptical

tischen Bewegung werden nun fortwährend geändert werden, und die von der betrachteten Masse beschriebene Bahn wird dadurch in eine Spirallinie übergehen, in welcher die betrachtete Masse endlich so weit von A fortgeführt wird, dass sie aus der Wirkungsphäre von A in die Wirkungsphäre von B gelangt, und so fort, nachdem sie eine Anzahl Spiralwindungen um B beschrieben hat, auch von B so weit fortgeführt wird, dass sie aus der Wirkungsphäre von B in die Wirkungsphäre von C gelangt. Auf diese Weise kann also eine elektromotorische Kraft ein Fortströmen der negativen Elektricität in der Richtung ABC bewirken, an welchem die positiven Massen in A, B, C keinen Antheil nehmen. Das Wesentliche dieser Betrachtung besteht darin, dass, sobald die elektromotorische Kraft zu wirken aufhört, die betrachtete Masse sogleich wieder nach den Kepler'schen Gesetzen in elliptischer Bahn um diejenige positive Masse sich bewegen wird, in deren Nähe sie sich gerade befindet, weil nach Wegfall der störenden Kraft keine weitere Aenderung der Elemente ihrer Centralbewegung Statt findet. Auch ersieht man leicht, dass in dieser wesentlichen Beziehung nichts geändert werden würde, wenn die positiven Massen in A, B, C ... gleichfalls beweglich angenommen und ausser der Centralkraft der negativen Massen, in deren Nähe sie sich befinden, der störenden Einwirkung der nämlichen elektromotorischen Kraft unterworfen würden, welche aber für diese positive Massen die entgegengesetzte Richtung, wie für die negative hätte. Es ergiebt sich daraus folgendes Resultat. Wenn die elektromotorische Kraft c auf die betrachtete negative Masse allein wirkte, so würde sie dieser Masse in der Richtung ABC während der Zeit t eine Geschwindigkeit ct ertheilen, mit welcher sich diese Masse, auch nachdem die Kraft c zu wirken aufgehört hätte, beharrlich in der Richtung ABC fortbewegen müsste. Unter Mitwirkung der Centralkräfte der positiven Massen in A, B, C .. aber wird zwar die elektromotorische Kraft c ebenfalls, so lange sie wirkt, ein Fortrücken der betrachteten Masse in der Richtung ABC bewirken, sobald die Kraft c aber zu wirken aufhört, wird auch dieses Fortrückeen aufhören, d. h. dieses Fortrücken der betrachteten Masse in der Richtung ABC geschieht dann nicht mit einer Geschwindigkeit, welche fortdauert, nachdem die Kraft zu wirken aufgehört, welche das Fortrücken hervorgebracht hat. Der Grund also, warum die betrachtete Masse in der Richtung ABC nicht weiter fortrückt, nachdem die elektromotorische Kraft zu wirken augehört hat, liegt darnach in den von den positiven Massen auf die betrachtete negative Masse ausgeübten Centralkräften. Das Wort *Widerstand* bezeichnet aber in der Theorie der galvanischen Kette wesentlich nichts Anderes, als das Faktum, dass die Fortbewegung der elektrischen Fluida im galvanishcen Strome der elektromotorischen Kraft proportional ist, d. h. aufhört, sobald die elektromotorische Kraft zu wirken aufhört. Es folgt also daraus, dass der Grund des Widerstands in den *Centralkräften* liegen kann, welche die im elektrischen Doppelstrome sich begegnenden positiven und negativen Massen wechselig auf einander ausüben. Es würde für weitere theoretische Untersuchung wichtig sein, aus diesen Grunde eine bestimmte und präcise Definition des Widerstands abzuleiten und die Beziehungen zu entwickeln, in welchen der nach seiner Wirkung definirte Widerstand dazu stehe. Es würde dabei hauptsächlich auf eine Bestimmung der Zeit ankommen, welche ein Theilchen braucht, um in seiner Spiralbahn von einer Windung um eine Centralmasse A zur entsprechenden Windung um die darauf folgende Centralmasse B zu gelangen. Dass aber solche Bestimmungen, auch wenn alle wesentlichen Elemente für die Rechnung gegeben sind, grosse Schwierigkeiten finden, zeigt die Theorie der Störungen in der Astronomie.

motion will be continuously changed, and the orbit of the considered mass will change into a spiral line, in which this mass will go so far from A, that it will pass from the sphere of influence of A to the sphere of influence of B, and so on, after it has described a number of spiral turns around B, so that it will be sent so far from B, that it will come into the sphere of influence of C. In this way an electromotive force can produce a current of the negative electricity along the direction ABC, in which the positive masses A, B, C do not take part. The essence of this consideration consists in the following, that as soon as the electromotive force stops acting, the considered mass will immediately move according to Kepler's laws in an elliptical orbit around the positive mass closest to it at this moment, due to the fact that with the absence of the disturbing force it does not happen any other change in the elements of its central motion. So we easily see that in this connection nothing would be changed if the positive masses A, B, C, ..., were also considered as mobile and were submitted, beyond the central force of the negative masses close to which they are found, to the perturbing influence of the electromotive force which, however, would have for these positive masses the opposite direction in relation to the negative masses. From this it follows the following result. When only the electromotive force c acted upon the considered negative mass, then this mass would acquire a velocity ct along the direction ABC during the time t, with which this mass, also after the [electromotive]force c had ceased to act, must continue to maintain its motion along the direction AB. However, with the joint contribution of the central forces exerted by the positive masses A, B, C, ..., the electromotive force c would likewise cause, while it acts, a continuous push upon the considered mass along the direction ABC, however, as soon as the [electromotive]force c stops acting, this continuous push will cease, that is, this continuous push upon the considered mass along the direction ABC does not happen then any longer with a lasting velocity, after the [electromotive] force ceased its action, which brought about the continuous push. Therefore, the reason why the considered mass does not continue its motion along the direction ABC, after the electromotive force ceased its action, is due to the central forces exerted by the positive masses upon the considered negative mass. But the word resistance designates in the theory of the galvanic circuit nothing more than the fact that the continuous motion of the electric fluids in galvanic currents is proportional to

> the electromotive force, that is, it ceases as soon as the electromotive force stops acting. Therefore it follows that the cause of the resistance can be in the central forces, which act mutually upon the positive and negative masses of the electric double current. It would be important for the broader theoretical analysis, to derive from it a definite and precise definition of the resistance and to develop its relation with the effect of the resistance. It would depend essentially in a determination of the time necessary for a particle to move in its spiral orbit from a turn around a central mass A to an equivalent turn around the following central mass B. However, the perturbation theory of the astronomy shows that these determinations, even when all relevant elements for the calculation are given, present great difficulties.

Weber's idea that the resistive force might be due to a Newtonian central force falling as $1/r^2$ does not seem feasible to us for two main reasons:[74] (1) The Newtonian forces are conservative, and (2) do not depend upon the velocities of the interacting bodies. The resistive force responsible for Ohm's law, on the other hand, is non-conservative and proportional to the drifting velocities of the mobile charges, acting against the motion of these charges.[75] Therefore, the origin of the resistive force must be sought somewhere else. This is a very difficult topic in physics. Even today there is no clear answer to this question.

At this point it was not yet clear to Weber what happened with the extra kinetic energy gained by the mobile charge during its transference between neighboring stationary masses due to the application of the external electromotive force. In Chapter 1.7 we will see that in 1875 Weber found a possible solution for this problem.

The idea of a double current had been previously supposed by Ørsted, Ampère and by Weber himself in 1846 and in 1855-57. From this time onwards Weber maintained the idea of a simple current. He kept it to the end of his life, as can be seen in his paper of 1871,[76] in his paper of 1875, *On the movement of electricity in bodies of molecular constitution*,[77] and in his posthumous work, *Determinations of electrodynamic measure: particularly in respect to the connection of the fundamental laws of electricity with the law of gravitation*.[78] We quote from this last work:[79]

---

74 [4, Appendix A: Wilhelm Weber and Surface Charges, p. 195–211].
75 [39].
76 [Web71, Sections 17–19, p. 281–294 of Weber's *Werke*] and [Web72, p. 132–144].
77 [Web75, p. 340–342].
78 [Web94a, p. 479–480 and 499] and [webb, p. 2–3 and 28].
79 [Web94a, p. 479–480] and [webb, p. 2–3].

It is interesting to pursue this present development of physical research further and indeed, one observes:

Firstly, that the theory of magnetism *can only be absorbed by the theory of electricity under the* assumption *of movable parts in the interior of all magnetic and magnetizable bodies, i.e., positive electrical molecules, which form* molecular currents around the negatively electrically charged ponderable molecules, *in the interior of all magnetizable bodies.*

Secondly, by the further consideration, that the theory of galvanism and of heat, in order likewise to be absorbed by the theory of electricity, must also presuppose movable parts in the interior of all galvanic conductors and heat conductors; that, however, these need by no means be different parts which produce in the interior of ponderable bodies, magnetism, and others which produce the motion of galvanism, and still others, whose movements produce heat; rather that the same parts, according to the difference of their motions, can produce magnetism, galvanism, and heat, sometimes together, sometimes separately, and that these moving parts in the interior of ponderable bodies are molecules of one electricity, which ought to be called positive electricity.

Thirdly, it is to be considered, that the movements of these positive electrical molecules around the negatively electrically charged ponderable molecules of bodies, either form closed orbits, or spiral orbits differing only slightly from circular orbits with periodically increasing and attenuating diameter, or spiral-shaped orbits with continuously increasing diameter, whereby they ultimately pass over into a ballistic trajectory, thus effecting the transfer of this electrical molecule from one ponderable molecule to another neighboring ponderable molecule, whereupon in part heat conduction, and in part galvanic currents in metallic conductors, are based.

Fourthly, and finally, it is furthermore to be considered, that by means of magnetic or electrodynamic induction from the outside, circular currents around the ponderable molecules of a body may be excited, or circular currents already in existence may be enhanced, weakened, or changed in direction.

In summary, from 1852 onwards he evolved from the conception of a double galvanic current to that of an electric current being composed by the flow of charges of a single sign.

In the next Chapter we consider another origin of his planetary model of the atom. This time it is related directly with his fundamental force law which he had proposed in 1846.

1.6 The Motion of Two Charged Particles Interacting According to Weber's Force Law

1.6.1 Weber's Force and Potential Energy

In 1846 Weber published his paper *Determinations of electrodynamic measure: concerning a universal law of electrical action*.[80] In this work he presented his fundamental force of interaction between two charged particles, which was the basis of all his subsequent works on electrodynamics. His force of interaction depends not only on the distance between the two interacting charges, but also upon their relative radial velocity and upon their relative radial acceleration. He derived it from Ampère's force between current elements, presented in its final form by Ampère in 1822.[81] In a current element ids, the current intensity is represented by i and the infinitesimal length of the element is represented by ds. The forces of Ampère and Weber, like the gravitational force of Newton (1687) and the electrostatic force of Coulomb (1785), represent actions at a distance.

In 1846 Weber obtained[82] the following force law between two charges e and e' separated by a distance r, moving relative to one another with a relative velocity dr/dt and relative radial acceleration d^2r/dt^2:

$$\frac{ee'}{r^2}\left(1 - \frac{a^2}{16}\frac{dr^2}{dt^2} + \frac{a^2}{8}r\frac{d^2r}{dt^2}\right) \ . \tag{1.1}$$

In 1852, he replaced the constant $a^2/16$ by $1/c^2$. In this way his force law could be expressed as:[83]

$$\frac{ee'}{r^2}\left(1 - \frac{1}{c^2}\frac{dr^2}{dt^2} + \frac{2r}{c^2}\frac{d^2r}{dt^2}\right) \ . \tag{1.2}$$

Weber's constant c is not the same as the modern light velocity in vacuum (which will be represent here by $c_L = 2.998 \times 10^8$ m/s), but $\sqrt{2}$ this last value.

---

80 [Web46] and [weba].
81 [Amp22b].
82 [Web46, Weber's *Werke*, p. 157] and [weba, p. 98].
83 [Web52b, Weber's *Werke*, p. 366].

That is, $c = \sqrt{2}c_L = 4.24 \times 10^8$ m/s. The Weber's vconstant c was first measured by Weber and R. Kohlrausch in 1855–1856. The final result they obtained can be expressed[84] as $c = 4.39 \times 10^8$ m/s.

There are many modern works discussing Weber's constant, Weber's force law,[85] light velocity and the wave theory of light.[86]

In 1848 Weber showed that his force law could be derived from a velocity-dependent potential energy.[87] In 1871 he expressed[88] his potential by the following expression:[89]

$$V = \frac{ee'}{r}\left(1 - \frac{1}{c^2}\frac{dr^2}{dt^2}\right). \tag{1.5}$$

The force given by Eq. (1.2) can be obtained from V through $-dV/dr$.

1.6.2 Weber's Introduction of an Inertial Mass for the Electric Fluids

In 1864 Weber introduced explicitly the inertial mass to a charged particle:[90]

---

84 [Web55, Weber's *Werke*, p. 594], [WK56, Weber's *Werke*, p. 605] (English translation in [WK03]), [KW57, Weber's *Werke*, p. 652] and [WK68].

85 In vectorial notation and in the International System of Units SI we can express Weber's force $\vec{F}$ exerted by charge e on charge e' as

$$\vec{F} = \frac{ee'}{4\pi\varepsilon_0}\frac{\hat{r}}{r^2}\left(1 - \frac{\dot{r}^2}{c^2} + \frac{2r\ddot{r}}{c^2}\right) = \frac{ee'}{4\pi\varepsilon_0}\frac{\hat{r}}{r^2}\left(1 - \frac{\dot{r}^2}{2c_L^2} + \frac{r\ddot{r}}{c_L^2}\right). \tag{1.3}$$

In this equation $\varepsilon_0 = 8.85 \times 10^{-12}$ C^{2}N^{-1}m^{-2} is called the permittivity of free space, $\hat{r}$ is the unit vector pointing from e to e', the relative radial velocity is represented by $\dot{r} \equiv dr/dt$, the relative radial acceleration is represented by $\ddot{r} \equiv d^2r/dt^2$ and $c_L = c/\sqrt{2} = 2.998 \times 10^8$ m/s is the light velocity in vacuum.

86 [Wie60, p. 107–113], [Wie67, p. 138–141], [Wie93], [Wie94], [1], [2, Section 11.2, p. 244–249], [63], [75], [Wie04] and [4, Section 1.4, p. 14–20].

87 [Web48a, p. 245 of Weber's *Werke*] with English translation in [Web66c].

88 [Web71, Section 4, p. 257 and Section 5, p. 259, footnote 1 of Weber's *Werke*] and [Web72, Section 4, p. 10 and Section 5, p. 11, footnote].

89 In the SI Weber's potential energy is written as

$$V = \frac{1}{4\pi\varepsilon_0}\frac{ee'}{r}\left(1 - \frac{\dot{r}^2}{c^2}\right) = \frac{1}{4\pi\varepsilon_0}\frac{ee'}{r}\left(1 - \frac{\dot{r}^2}{2c_L^2}\right). \tag{1.4}$$

90 [Web64, p. 181 of Weber's *Werke*]:

Nach unserer bisherigen Kenntniss muss zwar der Elektricität als einem Körper eine *Masse* zugeschrieben werden, und diese Masse übt auf eine andere ähnliche Masse eine *Kraft* aus; es fehlt aber noch an der Kenntniss des *Verhältnisses* jener Masse zu dieser Kraft. Die Kenntniss des *Verhältnisses* war nun auch nicht nöthig, so lange es sich um *Gleichgewicht-*

According to our present knowledge we must indeed attribute a mass to the electricity as a body, and this mass exerts a force to a similar mass; however, the knowledge about the ratio of this mass to this force is still missing. The knowledge of this ratio was not necessary, while we dealt with equilibrium phenomena or with constant motion, when it is enough the knowledge of the forces; the different amounts of electricity could be distinguished, instead of according to their masses, according to the amount of forces, which acted upon one and the same amount of electricity in the unit distance, and this last amount of electricity could be ascertained through the force, which arises in an equal amount of electricity in the unit distance. An amount of electricity determined in this way was really the so-called electrostatic unit of measure. However, when we are not dealing with the mere equilibrium or mere conservation of an already existing motion, but with an amount of electricity which receives new motion, which it did not have before, then it is necessary also the knowledge about the mass of the electricity which is put into motion, or the ratio of this mass to the force exerted by it at the electrostatic unit of measure in the unit distance, that is, a knowledge about the number of electrostatic units of measure, which go at the electricity with a unit of mass (milligram).

In Weber's sixth major Memoir published in 1871,[91] Weber considered the motion of two particles with charges e and e' interacting with one another according to his force law, Eq. (1.2). The inertial masses of the particles were expressed by, respectively, ε and ε'. In this case, he was considering a system of units for which the unit of mass is one milligram.[92] For positive charges, he

---

serscheinungen oder um *beharrliche Bewegungen* handelte, wo die Kenntniss der *Kräfte* genügte; die verschiedenen Elektricitätsmengen konnten dabei, statt nach ihren Massen, nach der Grösse der *Kräfte* unterschieden werden, die sie auf eine und dieselbe Elektricitätsmenge in der Einheit der Entfernung ausübten, und diese letztere Elektricitätsmenge konnte durch die *Kraft* bestimmt werden, die sie auf eine *gleiche* Elektricitätsmenge in der Einheit der Entfernung ausübte. Eine so bestimmte Elektricitätsmenge war nun wirklich die sogenannte *elektrostatische Maasseinheit*. Handelt es sich aber nicht um blosses *Gleichgewicht* oder um blosse *Erhaltung einer schon vorhandenen Bewegung*, sondern soll einer Elektricitätsmenge *neue Bewegung* ertheilt werden, welche sie vorher nicht besass, so reicht die blosse Kenntniss der Kräfte nicht aus, sondern es bedarf auch der Kenntniss der *Masse* der in Bewegung zu setzenden Elektricität, oder des *Verhältnisses* dieser Masse zu der von ihr auf die elektrostatische Maasseinheit in der Einheit der Entfernung ausgübten Kraft, d. i. der Kenntniss der *Zahl der elektrostatischen Maasseinheiten, welche auf die Masseneinheit* (Milligramm) *Elektricität gehen.*

91 [Web71], translated to English in [Web72].
92 [Web71, p. 250–251 of Weber's *Werke*] and [Web72, p. 2–3].

called the ratio charge to mass by a symbol $a > 0$, namely, $e/\varepsilon = e'/\varepsilon' = a$, while for negative charges he called the ratio charge to mass by a symbol $b < 0$, namely, $e/\varepsilon = e'/\varepsilon' = b$. The present constant a should not be confused with the constant a of Eq. (1.1). To our knowledge Weber was the first scientist to consider explicitly a charged particle with an inertial mass. Moreover, he considered that only experiment could decide about the ratio of a^2 to b^2. As the value of this ratio was not decided during his life, he left this ratio as an open question for the time being. This was an amazing insight which would prove extremely valuable in the future. For instance, nowadays we know the existence of protons and electrons. The magnitude of their charges is the same, while they have different masses. In Weber's language, the ratio a^2/b^2 would be different from 1 if we consider the value of a for a proton and the value of b for an electron.

Weber's specific words were as follows:[93]

1. *Electrical particles and Electrical Masses.*

> *Particles of positive and of negative electricity are denoted by the same letters, for instance by e or e' &c., but a positive or a negative value is assigned to e or e' ... according to whether it represents a particle of the positive or of the negative fluid.*
>
> *[...] For if we denote the masses of the particles e, e' (in the mechanical sense, according to which the unit of mass [1 milligramme] is determined by the mass of one ponderable body, and different masses are compared with each other in proportion to the reciprocals of the accelerations produced in them by the same force) by ε, ε', of which the values are always positive, we get for positive values of e, e',*
>
> $$\frac{e}{\varepsilon} = \frac{e'}{\varepsilon'} = a \;;$$
>
> *and for negative values of e, e',*
>
> $$\frac{e}{\varepsilon} = \frac{e'}{\varepsilon'} = b \;,$$
>
> *where a has a definite positive and b a definite negative value. Whether or not we have here $a^2 = b^2$, or what ratio a^2 bears to b^2, has not as yet been made out, any more than the numerical value of a or of b.*

---

[93] [Web71, p. 249–251 of Weber's *Werke*] and [Web72, p. 2–3].

1.6.3 Weber's Equation of Motion and His Critical Distance

In order to study the motion of two charged particles interacting with one another according to his force law, Weber considered separate situations. In the first situation the charges would be moving only along the straight line connecting them. In the second situation he allowed their motion to have components at right angles to the straight line connecting them, so that they could perform orbits around one another. Let us consider initially the first situation.[94]

Weber's force conserves energy, as shown by Weber himself. This means that the sum of the potential energy of the two particles with their kinetic energies is a constant in time if they are not interacting with other bodies. Weber defined r_0 as the distance between the charges when their relative radial velocity goes to zero, namely, when $dr/dt = 0$. His words were as follows:[95] "r_0 denoting the value of r for the moment when $dr/dt = u = 0$."

Weber also defined[96] a critical distance ϱ by:[97]

$$\varrho \equiv 2\left(\frac{1}{\varepsilon} + \frac{1}{\varepsilon'}\right)\frac{ee'}{c^2}. \tag{1.7}$$

The critical distance was called by Weber as a "molecular distance" in his paper of 1871.[98] Although Weber could not yet determine the exact value of this constant, he knew it would be extremely small and for this reason called it a molecular distance. Later on he also utilized[99] the expression "critical distance". Weber gave the name "molecular movements" when the particles were moving relative to one another separated by distances r smaller than ϱ, and called[100] "movements at a distance" when $r > \varrho$. He showed that no transition from movements at a distance to molecular movements could take place, so long as both particles were moving in consequence of their reciprocal action only. That is, if the movement did begin at $r < \varrho$, it would always remain inside this critical distance. On the other hand, if it began at $r > \varrho$, it would always

---

[94] For a complete solution of this problem in the International System of Units, see [18].
[95] [Web71, p. 269 of Weber's *Werke*] and [Web72, p. 120].
[96] [Web71, p. 269 of Weber's *Werke*] and [Web72, p. 120].
[97] The critical distance ϱ is defined in the *SI* as

$$\varrho \equiv \frac{ee'}{4\pi\varepsilon_0}\left(\frac{1}{\varepsilon} + \frac{1}{\varepsilon'}\right)\frac{2}{c^2} = \frac{ee'}{4\pi\varepsilon_0}\frac{\varepsilon+\varepsilon'}{\varepsilon\varepsilon'}\frac{1}{c_L^2}. \tag{1.6}$$

[98] [Web71, p. 298 of Weber's *Werke*] and [Web72, p. 148]. A discussion and references about this concept can be found in [Wie60, p. 140, 211, 212, 217 and 226], [18] and [70].
[99] [Web94a, p. 492 of Weber's *Werke*], with English translation in [webb, p. 18].
[100] [Web71, p. 268 and 271] and [Web72, p. 119 and 121].

remain outside this distance. This molecular or critical distance ϱ can have a positive or negative value, depending upon the sign of the product ee'.

With these definitions of r_0 and ϱ, and applying conservation of energy, Weber obtained the following equation of motion describing the interaction of two charged particles moving only along the straight line connecting them:[101]

$$\frac{dr^2}{dt^2} = \frac{r - r_0}{r - \varrho} \frac{\varrho}{r_0} c^2 \ . \tag{1.13}$$

In the sequence of his work, Weber considered the second situation. That is, he now allowed the motions of the particles to have components at right

---

[101] It is presented here the deduction of this equation in the International System of Units SI. We consider two particles along the x axis, moving along the straight line connecting them. According to the conservation of energy, the sum of Weber's potential energy V with the kinetic energy T is a constant in time for a system of two charges interacting with one another according to Weber's law. The charges of these two particles are represented by e and e', while their inertial masses are represented by ε and ε', respectively. The law of conservation of energy can be written as:

$$V + T = V_0 + T_0 \ , \tag{1.8}$$

where V_0 and T_0 are the initial values of V and T. The distances of charges e and e' to the origin of the coordinate system are represented by r_1 and r_2, respectively. The distance r between these two particles is given by $r = r_1 + r_2$, as both of them are considered along the x axis. The radial velocity $dr/dt = \dot{r}$ between these two particles is given by $\dot{r} = \dot{r}_1 + \dot{r}_2$. Let us consider an inertial system of reference in which the center of mass of the two particles is at the origin of coordinates. This yields $\varepsilon r_1 = \varepsilon' r_2$, or $r_1 = (\varepsilon'/\varepsilon)r_2$ and $\dot{r}_1 = (\varepsilon'/\varepsilon)\dot{r}_2$. Therefore,

$$\dot{r} = \dot{r}_1 + \dot{r}_2 = \frac{\varepsilon'}{\varepsilon}\dot{r}_2 + \dot{r}_2 = \frac{\varepsilon + \varepsilon'}{\varepsilon}\dot{r}_2 = \frac{\varepsilon + \varepsilon'}{\varepsilon'}\dot{r}_1 \ . \tag{1.9}$$

According to this equation, the kinetic energy of these two particles moving along the straight line connecting them can be written as

$$T = \frac{\varepsilon}{2}\dot{r}_1^2 + \frac{\varepsilon'}{2}\dot{r}_2^2 = \frac{\varepsilon}{2}\left(\frac{\varepsilon'}{\varepsilon + \varepsilon'}\right)^2\dot{r}^2 + \frac{\varepsilon'}{2}\left(\frac{\varepsilon}{\varepsilon + \varepsilon'}\right)^2\dot{r}^2 = \frac{\varepsilon\varepsilon'}{\varepsilon + \varepsilon'}\frac{\dot{r}^2}{2} \ . \tag{1.10}$$

Combining Eqs. (1.4) and (1.10) into Eq. (1.8) yields

$$\frac{ee'}{4\pi\varepsilon_0}\frac{1}{r}\left(1 - \frac{\dot{r}^2}{2c_L^2}\right) + \frac{\varepsilon\varepsilon'}{\varepsilon + \varepsilon'}\frac{\dot{r}^2}{2} = \frac{ee'}{4\pi\varepsilon_0}\frac{1}{r_0} \ . \tag{1.11}$$

The constant r_0 was defined by Weber as the value of r for which $\dot{r} = 0$.

Utilizing Weber's critical distance ϱ given by Eq. (1.6), this last equation can be put into the form

$$\dot{r}^2 = \frac{r - r_0}{r - \varrho}\frac{\varrho}{r_0}c^2 = \frac{r - r_0}{r - \varrho}\frac{\varrho}{r_0}2c_L^2 \ . \tag{1.12}$$

Eq. (1.12) is equivalent to Eq. (1.13).

angles to the straight line connecting them, so that they could perform orbits around one another.[102] In this second case, Weber obtained results analogous to the previous one, but now with motions in two- or three-dimensional space and no longer restricted along a straight line. For this more general situation, Weber[103] obtained the following equation of motion:[104]

$$\frac{uu}{cc} = \frac{r - r_0}{r - \varrho} \left(\frac{\varrho}{r_0} + \frac{r + r_0}{r} \frac{\alpha_0 \alpha_0}{cc} \right) . \tag{1.15}$$

Weber[105] put $u = dr/dt$ and "let α denote the difference of the two velocities which two electrical particles e and e', at a distance r from each other, possess in space in a direction perpendicular to the straight line r which joins them." That is, α was the tangential velocity of one particle relative to another and α_0 represented this relative tangential velocity when $r = r_0$ and $dr/dt = 0$.

1.6.4 Motion of Two Particles of the Same Kind

Weber first considered the situation in which the two interacting particles have charges of the same sign and are moving only along the straight line connecting them. The remarkable fact of Weber's electrodynamics is that, according to Eqs. (1.13) and (1.12), two particles with charges of the same sign can attract one another when they are in molecular movements, that is, when $0 \leq r < \varrho$. If two particles with charges of the same sign move relative to one another at distances r greater than ϱ, they will repel one another. This separates their motion[106] into two "states of aggregation." In the first state of aggregation, they will be always in molecular movements attracting one another and oscillating between $r = 0$ and $r = r_0 < \varrho$. The relative velocity dr/dt at $r = r_0$ will be null, while at $r = 0$ it will be $\pm c = \pm\sqrt{2}c_L$. Here the lower sign is valid just before their encounter and the upper sign just after it. In the second state of aggregation, they will be repelling one another but always remaining at a distance $r > \varrho > 0$. If they are initially moving away from one another, they will continue for ever this movement, until they are at an infinite distance from

---

102 A complete analytical solution of this problem in terms of elliptic functions already exists, [16].
103 [Web71, p. 272–274 of Weber's *Werke*] and [Web72, p. 123–125].
104 In the *SI* this equation would be written as

$$\frac{\dot{r}^2}{c^2} = \frac{\dot{r}^2}{2c_L^2} = \frac{r - r_0}{r - \varrho} \left(\frac{\varrho}{r_0} + \frac{r + r_0}{r} \frac{\alpha_0^2}{c^2} \right) = \frac{r - r_0}{r - \varrho} \left(\frac{\varrho}{r_0} + \frac{r + r_0}{r} \frac{\alpha_0^2}{2c_L^2} \right) . \tag{1.14}$$

105 [Web71, p. 272 of Weber's *Werke*] and [Web72, p. 123].
106 [Web71, p. 271 of Weber's *Werke*] and [Web72, p. 122].

one another, never coming back. If they are initially approaching one another, they will reach a smallest distance of separation $r_0 > \varrho$ for which $dr/dt = 0$. After this closest encounter, they will move away from one another due to their continuous repulsion, until reaching an infinite distance and never coming back again.

Weber then considered the second situation. That is, he now allowed the motions of the particles to have components at right angles to the straight line connecting them, so that they could perform orbits around one another. For the case of two particles with charges of the same sign, for instance, there would be once more two states of aggregation. In the first state of aggregation, they would attract one another when moving inside of sphere of diameter ϱ, that is, in molecular movements. In the second state of aggregation, they would repel one another when moving at a distance $r > \varrho$. This last solution was somewhat analogous to the Coulombian hyperbolic solution of two charged particles repelling one another with a smallest distance of closest approach, before being reflected to an infinite distance.[107] Weber expressed these two states of aggregation as follows, our words in square brackets:[108]

> *There results, in fact, in this case also, a distinction between two states of aggregation for two similar particles [that is, with charges of the same sign] – namely, a state of aggregation in which the two particles move in such a way as to return periodically into the same position relatively to each other [in a molecular movement for which $r < \varrho$], and a state of aggregation in which the two particles move so as to become always more and more distant from each other and never return to the same position. [In this second state of aggregation they are moving at a distance, in such a way that $r > \varrho$.] No transition from one state of aggregation to the other takes place so long as the two particles move only under the influence of their own reciprocal forces.*

The remarkable fact of two charges of the same sign attracting one another when they are very close to each other (that is, when $r < \varrho$, the critical distance) is a unique feature of Weber's electrodynamic force law. It will play an important role in his mature planetary model of the atom to be discussed in Chapter 1.10.

---

107 [Sym71, Section 3.16].
108 [Web71, p. 274 of Weber's *Werke*] and [Web72, p. 125].

1.6.5 Motion of Two Dissimilar Electrical Particles

Weber did not analyze explicitly the motion of two charges of opposite sign moving along the straight line connecting them. In any event, this solution can be worked out[109] from the equation of motion which he obtained, Eq. (1.13). For charges of opposite sign, the critical distance ϱ is negative. Therefore, they would always attract one another.

There is a solution for this problem of two charges of opposite sign moving along the straight line connecting them in which they would oscillate between $r = 0$ and $r = r_0$, where r_0 is the value of their distance r for which $dr/dt = 0$. This is somewhat analogous to the Coulombian case, with the difference that the relative velocity at $r = 0$ will be $\pm c = \pm\sqrt{2}c_L$, instead of the Coulombian value $\dot{r} \to \pm\infty$ at $r \to 0$.

There is another solution for this problem for which the two charges would collide at $r = 0$ with a relative velocity $\dot{r} = -c = -\sqrt{2}c_L$, moving away from one another after this close encounter. They would never meet again. When they were at an infinite distance from each other, they would still be moving away from one another with a finite relative velocity. This is analogous to the Coulombian solution, except for the finite velocity during the collision.

Weber considered explicitly the motion of two dissimilar electrical particles (that is, for which $ee' < 0$) moving in directions at right angles to the straight line connecting them.[110] For two particles with charges of opposite sign, there was a particular solution equal to the Newtonian or Coulombian case. That is, a rotation of the two particles about each other while remaining at a constant distance from each other during this rotation. This kind of solution of a circular orbit was not possible for two charges of the same sign interacting with one another according to Weber's law.

However, when the distance r between the particles of opposite signs orbiting around one another changed during their motion, Weber found a solution according to his force law in which the distances between the two particles would always be confined between two limiting radii for which $dr/dt = 0$. In 1875 he[111] represented these two limiting radii by r_0 and r^0, with $r^0 \geq r_0$. This is somewhat similar to the Newtonian or Coulombian solutions of a Keplerian elliptical orbit in which the two bodies oscillate between the smallest r_0 and the largest distance r^0, while orbiting around one another. In Weber's words,[112] "the two particles must always remain in *oscillatory motion relatively to each*

---

109 See [18].
110 [Web71, Section 16, p. 279–281 of Weber's *Werke*] and cite[p. 130–132]weber72.
111 [Web75, p. 341 of Weber's *Werke*].
112 [Web71, p. 281 of Weber's *Werke*] and [Web72, p. 132].

other within the indicated limits." However, according to Weber's fundamental force law, there would be no longer the exact Coulombian or Newtonian solution of a *closed* Keplerian elliptical orbit. According to Weber's law, in this case the axis of the ellipse would precess while the particles were orbiting around each other between the limiting radii.

There are many works discussing mathematically the precession of the perihelion utilizing Weber's law applied to electromagnetism and gravitation.[113]

These states of aggregation of two charges of the same sign, or of two charges of opposite sign, are extremely similar to the atomic model of Rutherford and Bohr, which would be developed only 40 years later.

1.7 Weber's Speculations about the Conduction of Electricity and Heat in Conductors

In this Chapter we present the speculations of Wilhelm Weber related to the conduction of heat and electricity in conductors.[114]

Since the beginning of the XIXth century it was known that a metallic conductor is heated when an electric current flows through it. Around the middle of the XIXth century it was accepted the notion (which had been already discussed before the middle of the century), that heat is related to the motion of small particles inside a body, instead of being considered a material substance, like the so-called caloric. In Weber's time, it was known that heat is related to the internal kinetic energy of the body, that is, is related to motion. A greater temperature for a body means that it has a greater internal random kinetic energy. According to this point of view, a transference of heat is nothing more than a transference of motion or of kinetic energy. But at that point it had not yet been completely clarified the question of what inside a body is in motion: the ponderable matter (the molecules composing the body), the contained ether inside the body (sometimes called the heat medium), or the electric particles composing the body. It was also not clear which kind of motion was related to heat. It could be an oscillatory motion or a rotational motion, for instance.

The law of generation of heat by the current was established by James Prescott Joule (1818–1889), Alexandre Edmond Becquerel[115] (1820–1891) and Heinrich Friedrich Emil Lenz (1804–1865) in the 1840's. Joule first established

---

113 [see] with German translation in [See24], [Tis72], [Zol76, p. xi–xii], [Zol83, p. 126–128], [ser], [Tis95], [Ger98], [Ger17], [Sch25] with English translation in [Sch95], [Nor65, p. 46], [Whi73, p. 207–208], [Eby77], [11], [16], [1, Sections 7.1 and 7.5], [2] etc.
114 [Wie60, p. 182–197], [Wie67, p. 157–161 and p. 169–177], [Wie88] and [Wie07].
115 Father of Henri Becquerel (1852–1908), one of the discoverers of radioactivity.

experimentally that the heat produced in a given time is proportional to the square of the current, to the resistance of the conductor, and to the time during which the constant current flows through the conductor.

Weber wanted to explain the existence of permanent molecular currents and also the generation of heat in conductors based on a single principle, namely, on the motion of charged particles. Due to the mechanical theory of heat, it was possible to suppose that heat was related to the motion of small particles inside the body. The transformation of electrical energy into heat suggested a close connection between this motion and the motion of the electrical particles composing a current. This possibility was also based on the discovery that a good conductor of heat is also a good conductor of electricity. The laws relating the thermal and electrical conductivities of metals, and their connections with the temperature, were obtained by Gustav Wiedemann (1826–1899), Rudolph Franz (1827–1902) and Ludwig Lorenz (1829–1891) in the 1850's and 1870's.

Weber first exposed his ideas in this direction in a work of 1862 related to galvanometry.[116] He extended his analysis in 1875 in a work discussing the motion of electricity in bodies with molecular constitution.[117] In Chapter 1.5 it was shown that in 1852 Weber had began to consider the Ampèrian molecular current as a positive charge orbiting around a stationary negative charge (the signs might also be reverted according to Weber) in a Keplerian elliptical orbit. In 1862 and 1875 this stationary negative charge was supposed to be merged with the ponderable molecule of the conductor. As regards a galvanic current, Weber considered that the application of an external electromotive force would transform this elliptical orbit into a spiral line. The mobile charge would be orbiting a specific stationary charge with increasing radius, until it would fall into the sphere of influence of another stationary charge along the direction of the electromotive force, beginning to orbit around it. This process would continue while the electromotive force were still active. In order to clarify the transformation of electrical energy into thermal energy, Weber considered in 1875 that the mobile charge would arrive at the second stationary charge with a higher translational kinetic energy than it had when it left the first stationary charge, due to the applied electromotive force along the direction of the conductor. It would then lose this extra translational kinetic energy to the next stationary charge. Therefore, on average, the mobile charge would then move along the direction of the applied electromotive force with a constant velocity. The lost translational kinetic energy of the mobile charge would had been transformed into thermal energy of the molecule. In 1875 he considered that

---

116 [Web62, Section 33, p. 91–96 of Weber's *Werke*].
117 [Web75, Sections 4 to 9, p. 334–353 of Weber's *Werke*].

this lost translational kinetic energy would be gained by the particle orbiting around the molecule. That is, the lost translational kinetic energy would be transformed into a rotational kinetic energy of the orbiting particle. And this greater orbital or rotational kinetic energy would be equivalent to a greater thermal energy of the body. In other words, the heating of a metallic conductor due the the flow of a current would be equivalent to a higher rotational motion of the charged particles composing the Ampèrian molecular currents.

In this work Weber did not consider the possibility that the greater temperature of the body might be related with a vibration of the ponderable atoms or molecules as a whole. He did not imagine that the heat might be related with a random motion of the ponderable molecules, that is, independent of the random motion of the electrified particles of the body.

Let us consider here some specific examples of his reasonings. In Section 4 of his paper of 1875, Weber considered the possibility of relating three phenomena with internal motions of a body, namely, galvanic currents, magnetic phenomena and heat. He speculated that all these phenomena might be related to internal motions of the same corpuscles, namely, charged particles. He considered that galvanic and electrodynamic phenomena might be due to the translational motions of these charged particles. As regards magnetic and diamagnetic phenomena, he supposed them to be due to the rotational motions of these charged particles around the ponderable particles of the body, like a planetary Ampèrian molecular current. According to the mechanical theory of heat, he considered in this work that also the heat might be related with the rotational motion of these charged particles around the ponderable particles of the body.

In the next Section of his paper, Weber considered the identity between the kinetic energy generated by the electromotive force, with the heat generated in the conductor by the galvanic current. If a conductor had no resistance, the application of an external electromotive force would increase indefinitely the translational kinetic energy of the mobile charges. But this does not happen in a resistive conductor, according to Ohm's law. As the current remains constant for a constant applied electromotive force, this means that on average the translational kinetic energy of the mobile particles do not increase while they move along the resistive circuit, despite the application of the electromotive force. Weber's conclusion was that the work generated by the external electromotive force must be transformed into another kind of motion, in accordance with the conservation of energy. He then mentioned that in magnetic and diamagnetic phenomena the electric particles could be supposed to orbit around the ponderable molecules of the conductor. He returned once more to his model of 1852 in which he considered this planetary Ampèrian molecular current and a sequence of ponderable molecules A, B, C etc. along a straight line. The positive electrical particles would orbit around a ponderable molecule merged with a negative charge.[118] With the application of the external electromotive force, the positive charged particle would be accelerated along this direction, while orbiting around a negative ponderable molecule A. It then would go into an spiral orbit, until it reached the sphere of influence of the next negative ponderable molecule B, beginning to orbit around it. It would arrive at this second ponderable molecule B with a higher translational kinetic energy than the translational kinetic energy it had when it left the first ponderable molecule A. It would then lose its extra translational kinetic energy to this ponderable molecule B, departing from it with the same translational kinetic energy that it had when it left molecule A. This process would repeat itself at molecules C, D etc. The novelty in this work of 1875 is that he considered the possibility that this extra translational kinetic energy which the charged particle lost at B might be transformed into a rotational kinetic energy of other charged particles which were orbiting around molecule B, being responsible for its magnetic and diamagnetic phenomena. The same transference or transformation of energy would happen at C, D etc. And this greater rotational energy of the charged particles orbiting around B, C, D etc. would represent the thermal kinetic energy. That is, the temperature increase would correspond in this Weberian model to an increase of the rotational kinetic energy of the charged particles orbiting around the ponderable molecules of the body.

Weber spoke of a ballistic trajectory or projectile motion (*Wurfbewegung*) when a positive electrical particle was emitted from one negative ponderable molecule around which it was orbiting, being accelerated towards another negative ponderable molecule, due to the application of an external electromotive force. He also spoke of an average distance between the molecules (*mittlerer Molekularabstand*).[119]

We present here some of Weber's own words connecting heat and the kinetic energy of his planetary molecular currents:[120]

> *This increase of kinetic energy of the electrical particles contained in a conductor while a current traverses it, follows therefore as a necessary consequence of the action of the electromotive force upon the particles, while these particles, as the result of the current, move onward in the direction of this force.*
>
> *This theoretical conclusion receives, not indeed a direct, but an in-*

---

118 [Web75, p. 348 of Weber's *Werke*].
119 [Web75, p. 349 of Weber's *Werke*].
120 [Web71, p. 292–294 of Weber's *Werke*] and [Web72, p. 143–144].

> *direct confirmation from experiment, inasmuch as an increase of thermal energy is observed in the conductor while current traverses it. And this observed increase of thermal energy in the conductor is equal to the calculated increase of the kinetic energy of the electrical particles in the Ampèrian molecular currents of the conductor.*
>
> *Now the thermal energy of a body is a kinetic energy resulting from movements in the interior of the body, which are therefore inaccessible to direct observation. In like manner, the kinetic energy belonging to the electrical particles in the Ampèrian molecular currents in a conductor is a kinetic energy which results from movements taking place in the interior of the conductor, and therefore inaccessible to direct observation. [...]*
>
> *Hence it follows as a consequence that, if in conductors all the electrical particles exist in the state of aggregation corresponding to Ampèrian molecular currents, the observed increase in the thermal energy of a conductor, during the passage of a current through it, must result immediately from the increase of the kinetic energy of the electrical particles constituting the Ampèrian currents; that is to say, the thermal energy imparted to the conductor by the current must be kinetic energy due to motions in the interior of the conductor, and must in fact consist in an increase in the strength of the Ampèrian currents formed by the electrical particles in the conductor.*

Many scientists recognized the role of Wilhelm Weber as the initiator of the modern theory of electric conduction in metals.[121] To give an example, we quote here the words of Drude published originally in 1905:[122]

> *The suggestion that electric conduction in metals is essentially the same as in electrolytes, that is to say, that it is effected by the movement of small charged particles, was first made by W. Weber,*[123] *who made use of it to derive Ohm's law.*

---

[121] [Wie07].
[122] [Dru05, p. 253].
[123] [Note by Drude, referring to [Web62, p. 91 of Weber's *Werke*], [Web71], with English translation in [Web72], [Web75] and [Web94a], with English translation in [webb]:] W. Weber. "Gesammelte Werke," 4, p. 91, 1862; p. 247, 1871; p. 312, 1875; p. 479.

1.8 Weber's Speculations about the Conduction of Heat in Insulators

In Weber's article of 1875, he tried to understand the difference between a conductor and an insulator.[124] Both of them conduct heat, but only the former conducts electricity. He had explained the transformation of electrical energy into thermal energy in conductors supposing that the extra translational kinetic energy gained by the mobile charged particles composing a galvanic current was transformed into a rotational kinetic energy of the mobile charged particles orbiting around the ponderable molecules of the conductor. This extra translational kinetic energy would be gained by each mobile charged particle due to the applied electromotive force, while this mobile charged particle would be transferred from the orbit around one oppositely charged molecule to the orbit around another oppositely charged molecule along the direction of the applied electromotive force. Weber called this kind of propagation of heat in metallic conductors by the name *propagation of heat through emission (Wärmeverbreitung durch Emission)* or simply *conduction of heat (Wärmeleitung)*.[125]

But how could he explain the conduction or transference of heat inside insulators? Or the propagation of heat through empty space? No electric currents can flow through these materials. In these cases, he talked about a second kind of heat propagation, which he called *heat propagation through radiation (Wärmeverbreitung durch Strahlung)* or simply *radiation of heat (Wärmestrahlung)*.[126]

As was seen in Chapter 1.6, for two charges of opposite sign orbiting around one another according to Weber's force law, he found a solution[127] for which the distance between the two particles would be oscillating between two limiting radii, namely, r_0 and r^0, with $r^0 \geq r_0$. Only in a very special case these two radii would coincide with one another giving rise to closed circular orbits. The values of these two limiting radii would depend upon the intrinsic properties of the atomic pair, namely, their electrical charges e and e', and their inertial masses ε and ε'. The values of these two radii would also depend upon the initial distance of separation of the two charges composing the atomic pair, upon their initial velocities along the direction connecting them and also upon their initial velocities orthogonal to this direction. In his work of 1875 Weber was following his earlier papers of 1852 and 1871 in which he supposed a planetary Ampèrian molecular current with the positive particle orbiting around a negative particle associated with a ponderable molecule. He also considered the mass of the

---

124 [Web75, Sections 6 and 7, p. 339–343 of Weber's *Werke*].
125 [Web75, p. 343 of Weber's *Werke*].
126 [Web75, p. 343 of Weber's *Werke*].
127 [Web75, p. 341 of Weber's *Werke*].

positive particle as negligible in comparison with the joint mass of the negative charge and its associated ponderable molecule. He thought an insulator as being composed of an array of ponderable molecules surrounded by planetary Ampèrian molecular currents.

Depending upon these properties, Weber was able to distinguish between two classes of bodies, which he called conductors and insulators.[128] Like Weber, let us call r^0 the largest distance between a positive particle and a negative molecule, with the positive particle orbiting around the negative molecule. As the mass of the negative molecule was considered much larger than that of the positive particle, the negative molecule can be considered at rest. According to Weber, conductors would be those bodies for which r^0 would be so large as to reach the sphere of influence of a neighboring negative molecule. In this case, it would be possible the transference of the positive particle from the sphere of influence of the first negative molecule to the sphere of influence of the second negative molecule. Insulators, on the other hand, would be those bodies for which the largest distance r^0 would not be large enough to reach the sphere of influence of any neighboring molecule. Consequently, it would not be possible the transference of a positive charge orbiting around a negative molecule to any other negative molecule, even with the application of an external electromotive force. Weber formulated this distinction mathematically.

In order to explain the conduction of heat in insulators and through space, Weber considered the existence of a fine medium between the molecules composing a conductor and also in space. He gave to this medium the name heat ether (*Wärmeäther*) or light ether (*Lichtäther*).[129] As regards this ether, Weber always supposed it as having a corpuscular or granular structure. Moreover, he considered it as being composed of positive and negative electrical particles. For instance, in his paper of 1846 he made the following comments:[130]

> *The idea of the existence of such a transmitting medium is already found in the idea of the all-pervasive neutral electrical fluid, and even if this neutral fluid, apart from conductors, has up to now almost entirely evaded the physicists' observations, nevertheless there is now hope that we can succeed in gaining more direct elucidation of this all-pervasive fluid in several new ways. Perhaps in other bodies, apart from conductors, no currents appear, but only vibrations, which can be observed more precisely for the first time with the methods discussed in Section 16. Further, I need only recall*

---

128 [Web75, p. 341 of Weber's *Werke*].
129 [Web62, p. 94–96 of Weber's *Werke*].
130 [Web46, p. 213–214] and [weba, p. 141–142].

Faraday's latest discovery of the influence of electrical currents on light vibrations, which make it not improbable, that the all-pervasive neutral electrical medium is itself that all-pervasive ether, which creates and propagates light vibrations, or that at least the two are so intimately interconnected, that observations of light vibrations may be able to explain the behavior of the neutral electrical medium.

Like his contemporaries, Weber proposed in 1875 that the light and heat radiations did propagate in the form of waves in this material ether. Although the physicists of that time had no clear proofs of the properties of this medium, Weber believed that we could have an idea of its behavior by supposing that this medium was composed of electrical particles. According to Weber, all bodies were composed of positive and negative electrical particles in different arrangements and states of motion. Weber tried to derive all physical phenomena based on this hypothesis. This included electrostatic, magnetostatic, galvanic, electrodynamic and thermal phenomena. In 1875 Weber proposed to extend this range of phenomena in order to include also optical phenomena and heat radiation. Weber's words:[131]

It is generally known that for the heat propagation through radiation in empty space or in insulators it is true the same as for light radiation, namely, that it is mediated through wave propagation, which presupposes the existence of a medium for the propagation of waves. Up to now we have tried to discover the property of this medium from the laws of wave motion, as they were found through observations of the light phenomena; now if this medium is composed of electricity, and if we had a better knowledge of its constitution, then it would be possible to develop the wave motion and to explain the light phenomena beginning with the fundamental laws of electric action, which in fact has been tried from different perspectives, but it would carry us too far away to go into it here.

---

[131] [Web75, p. 343 of Weber's *Werke*]:
Für diese Wärmeverbreitung durch Strahlung im leeren Raume oder in Isolatoren gilt bekanntlich dasselbe wie für die Lichtstrahlung, nämlich dass sie durch Wellenfortpflanzung vermittelt wird, was die Existenz eines wellenfortpflanzenden Mediums voraussetzt. Die Beschaffenheit dieses Medums hat man bisher aus den Gesetzen der Wellenbewegungen, wie sie aus den Beobachtungen der Lichterscheinungen gefunden worden, kennen zu lernen gesucht; bestände nun aber dieses Medium aus Elektricität, und besässe man nähere Kenntniss von seiner Konstitution, so würde es möglich sein, aus dem Grundgesetze der elektrischen Wirkung die Gesetze jener Wellenbewegungen zu entwickeln und die Lichterscheinungen daraus zu erklären, was auch wirklich auf verschiedene Weise versucht worden ist, worauf aber näher einzugehen hier zu weit führen würde.

In 1878 and in his posthumous work, Weber maintained the idea of a granular or corpuscular ether composed of electrical particles. But now he assumed that it would be composed of positive electrical molecules only.[132]

1.9 Optical Properties of Weber's Planetary Model of the Atom

Weber always defended a wave theory of light. For instance, in his joint book with his brother, the physiologist Ernst Heinrich Weber, Weber compared the wave theory and Newton's emanation theory, showing the advantages of a wave propagation through an ether.[133]

He discussed alternating currents in Section 16 of his first major Memoir of 1846. He distinguished the progressive galvanic current from this alternating current, which in very short sequential time intervals constantly changes its direction. He then advanced a bold hypothesis that incident light waves might create electrical vibrations upon the electrical fluids of a material substance! To our knowledge this was the first time he pointed out a possible connection between light and electricity:[134]

> *Since the progressive motion of electricity occurs so abundantly in Nature, it is not obvious why, given such great mobility occasional conditions should not also occur, which favor a vibrating movement. If, e. g., light undulations exert an effect on the electrical fluids, and have the power to disturb their equilibrium, it would certainly be expected that these effects of light undulations would be structured in time with the same periodicity as the light undulations themselves, so that the result would consist of an electrical vibration, which, however, we are unable to discover with our instruments.*

The electromagnetic influence upon optical phenomena was known since 1845, when Faraday discovered the magnetic rotation of the plane of polarization of light.[135] He observed this rotation for light traversing a piece of heavy glass immersed in a strong magnetic field along the direction of light propagation, or with the glass surrounded by a galvanic current flowing along an helix surrounding the glass, that is, with the current almost in a plane

---

132 [Web78, p. 383 and 394–395], [Web94a, p. 480, 489–491, 506, 516–518 and 524–525 of Weber's *Werke*] and [webb, p. 3, 15–17, 35, 48–51 and 57–58].
133 [WW93, Paragraphs 306 to 313 of Weber's *Werke*].
134 [Web46, p. 124 of Weber's *Werke*] and [weba, p. 76].
135 [Far65b, Series XIX, articles 2146–2242].

orthogonal to the light beam. Weber was well aware of this discovery and mentioned it in his paper of 1846, when advancing the suggestion that the ether believed to propagate light vibrations might be a neutral electrical medium, as seen in the quotation presented in Chapter 1.8.

Figure 1.6:
Carl Neumann (1832–1925)
(Wikimedia)

Carl Neumann (1832–1925), the son of Franz Neumann (1798–1895), treated mathematically the rotation of the plane of polarization of light by magnetism from the point of view of Weber's electrodynamics in his Dissertation of 1858.[136] Five years later he presented a more detailed account of his theory.[137] He proposed an interaction between ether particles and the molecules of the body depending upon external magnetic forces. But these forces would act only upon mobile ether particles which had been previously excited, but not upon stationary ether particles. He supposed this force to be produced by Ampèrian molecular currents induced in the body by these magnetic forces, in analogy with Weber's explanation for diamagnetism. However, it must be emphasized

---
136 [Neu58].
137 [Neu63].

here that Neumann utilized an *ideal* model of an Ampèrian molecular current, that is, based upon a continuous current flow around the molecule, like the rings of Saturn. These induced molecular currents would act upon the ether particles according to Weber's force law. This interaction would be analogous to the mutual interaction of two electrical currents. Neumann's theory was a first tentative to apply Weber's law to optical phenomena.[138]

In 1862 Weber postulated the excitation of heat- or light-waves through molecular currents.[139] Neumann's *ideal* model of an Ampèrian molecular current, on the other hand, could not excite these waves in the ether, as just mentioned. In this work of 1862, on the other hand, Weber presented once more a model for *discrete* or *corpuscular* Ampèrian molecular currents similar to the model which he had presented in 1852 and which we discussed in Section 1.3.5. This crucial change made it possible the excitation in the ether of light-waves through molecular currents. The only difference of this model as regards Weber's model of 1852, is that now Weber reversed the signs of the mobile and stationary electric charges. In this work of 1862 Weber endowed his planetary model of the atom with optical properties, namely, the possible production of light waves through the ether.

He made the following comments when discussing Neumann's work related to Faraday's rotation:[140]

---

138 [Wie60, p. 194–195].
139 [Web62, Weber's *Werke*, p. 94–96].
140 [Web62, Weber's *Werke*, p. 95]:

Zwar hat Neumann nach seinen Prämissen gefunden, dass keine Einwirkung elektrischer Molekularströme auf *ruhende Aethertheilchen* statt finden könne; es ist aber dabei zu beachten, dass diese Prämissen, dem Zwecke der Neumann'schen Untersuchung gemäss, welcher auf die Einwirkung der Molekularströme auf die schon vorhandenen mitten zwischen den Molekulen durch den Aether fortgepflanzten Wellenzüge beschränkt war, sich zwar auf Wirkungen der Molekularströme in sehr kleinen Entfernungen bezogen, doch aber noch immer die Zulassung einer *idealen* Vorstellung von den Molekularströmen gestatteten, wonach dieselben als eine *Superposition entgegengesetzt gleicher Ströme positiver und negativer Elektricität* betrachtet werden, was aber offenbar nicht gestattet ist, wenn es sich um die Erregung neuer Wellenzüge durch die elektrischen Molekularströme handelt, welche nur in der an die Molekularströme *unmittelbar angrenzenden Aetherschicht* Statt finden kann. Für diese Aetherschicht dürfen die in entgegengesetzter Richtung sich bewegenden positiven und negativen elektrischen Theilchen nicht mehr als zusammenfallend betrachtet werden. Denkt man sich dann also z. B. das negative Fluidum mit dem Molekule als fest verbunden, und das positive Fluidum allein in Molekularströmung begriffen, oder umgekehrt (eine Vorstellungswese, welche sich dadurch empfiehlt, dass sie mit der Beharrung der Molekularströme ohne elektromotorische Kräfte bestehen kann) so leuchtet ein, dass die Verschiedenheit in Lage und Verhalten beider elektrischen Fluida im Bereiche des Moleküls zwar schon bei sehr geringen Entfernungen (wie sie Neumann betrachtet) nicht mehr beachtet zu werden braucht, worauf die Zulässigkeit jener *idealen* Vorstellung von den Molekularströmen beruht, dass sie doch aber für die *unmittelbar an-*

> *Neumann found, according to his assumptions, that there could be no action of electric molecular currents upon stationary ether particles; however, it should be observed that these assumptions were in agreement with Neumann's goals, which were limited to the action of the molecular currents upon the wave trains propagating in the ether and already existing in the middle of the molecules, indeed to the actions of the molecular currents attaining to very small distances, it afforded the admission of an ideal conception of molecular currents, in which these were considered as a superposition of opposite and equal currents of positive and negative electricity, but which apparently is not appropriate for the production of new wave trains through the electric molecular currents, which can only happen in the immediately adjoining layer between the ether and the molecular currents. For these ether boundaries the considered electric particles moving in opposite directions should no longer be considered as coincident. When we suppose, for instance, the negative fluid as rigidly connected with the molecule, and consider only the positive fluid in molecular current, or vice versa (a conception which recommends itself, as it is consistent with the persistence of the molecular currents without electromotive forces), it is then clear, that the difference in position and behavior of both electric fluids in the domain of the molecule, indeed already by very small distances (as Neumann considered them) does not need any longer to be considered, based upon the admissibility of that ideal conception about the molecular currents, that it, however, for the immediately adjoining ether layer can have a significance, especially when the electric fluids composing the molecular currents were not continuous and evenly distributed around the molecule.*

Carl Neumann had an image of the molecular currents as composed of both electric fluids moving in opposite directions in closed orbits around the molecule, in analogy with the rings of Saturn. But this picture was not appropriate for the production of new waves through the ether. As we can see from this quotation, Weber modified Neumann's conception. The continuous distribution of positive and negative mobile charges moving around the molecule were now considered as concentrated in particles, like the Moon orbiting around

---

grenzende Aetherschicht von Bedeutung sein kann, zumal wenn das in Molekularströmung befindliche elektrische Fluidum nicht stetig und gleichförmig um das Molekule vertheilt wäre.

the Earth. That is, Weber's transformed Ampère's molecular currents into a planetary system!

As we had seen, in 1852 he had already a similar idea, but at that time with the positive charge considered as stationary with the molecule, while the negative charge orbited the positive molecule. In this work of 1862 he reversed the signs of the charges. At that time it was not yet possible to decide which sign of the charge should be connected with Ampère's molecular current.

In the sequence of this work of 1862, Weber even pointed out that the orbital frequency of the charged particles of his planetary model should be identical with the frequency of the excited heat- or light-waves. The relevant quotation runs as follows:[141]

> *When a perturbation of the equilibrium in the immediate border of the ether and, consequently, a production of an ether-wave, really takes place, then it is clear that it will repeat itself in each orbit of the electricity around the molecule, in such a way that the period of the wave must be identical with the period of the orbit of the electric particle which is in molecular current.*

Weber did not discuss the consequences of energy conservation in this production of heat- or light-waves by his planetary molecular current.

Weber also believed that it would be possible to utilize the optical properties of his planetary model in order to obtain information about the internal constitution of molecules. The wavelengths of the emitted light, in particular, might yield the key to draw conclusions from the electrical molecular processes:[142]

> *However, the wavelength of the wave train emitted by glowing molecules is well known from optical experiments; therefore, if the supposed relation between electrical molecular currents and the light ether, according to Neumann's ideas, are corroborated, then it would*

---

141 [Web62, p. 95 of Weber's *Werke*]:
Findet dann aber wirklich eine Störung des Gleichgewichts in der *unmittelbar angrenzenden Aetherschicht*, folglich eine Erregung von Aetherwellen, Statt, so leuchtet ein, dass dieselbe mit jedem Umlauf der Elektricität um das Molekul sich wiederholen, also die *Wellendauer* mit der *Umlaufszeit der elektrischen Theilchen* im Molekularstrome übereinstimmen muss.

142 [Web62, p. 95–96 of Weber's *Werke*]:
Bei *leuchtenden Molekulen* ist aber die Wellendauer der von ihnen ausgesandten Wellenzüge aus optischen Versuchen genau bekannt; es würde also, wenn die angenommene Relation zwischen elektrischen Molekularströmen und dem Lichtäther, nach Neumann's Idee, sich bestätigte, hiernach möglich werden, aus optischen Versuchen über das Verhalten der die Molekularströme bildenden Elektricität nähere Auskunft zu erhalten.

> be possible to obtain, from optical experiments, a better information
> about the behavior of the electricity generating a molecular current.

In 1876 Zöllner (cf. Figure 3.2, S. 162) advanced the other side of this reasoning, namely, to utilize the internal properties of a planetary model in order to derive the spectral lines of the chemical elements! The relevant quotation runs as follows:[143]

> The laws developed by Weber about the oscillations of an atomic pair will probably lead to an analytical determination of the number and position of the spectral lines of the chemical elements and their connections with the atomic weights.

This is a remarkable passage indicating a possible theoretical explanation of the known spectral lines of the elements. At that moment there was no known detailed explanation for these spectral lines. The spectral analysis of the chemical elements had been developed by R. W. Bunsen (1811–1899) and G. R. Kirchhoff (1824–1887) in 1859. The full quantitative understanding of the specific spectral series for each atom was attained only in the XXth century. In any event, it is amazing how far ahead of their time were Weber and Zöllner with these reasonings.

In this quotation, Zöllner was referring to Weber's work of 1871. Weber had estimated the period of oscillation of two charges of the same sign orbiting around one another separated by distances smaller than the critical distance. That is, separated by a distance r such that $r < \varrho$. He found that this period of oscillation was approximately between $2r_0/c$ and $4r_0/c$. He then made the following comment, trying to connect this period of oscillation with that of visible light:[144]

> If we put $c = 439450 \cdot 10^6$ millimetre/second, it follows from this last determination that the value of ϱ must lie approximately between $1/4000$ and $1/8000$ of a millimetre in order that these oscillations may be equal in rapidity to those of light.

As discussed by Hecht, it was with this model that Weber first attempted to find the basis for the production of oscillations of the frequency of light.[145]

---

143 [Zol76, Vorrede, p. XXI]:
Die von Ihnen [Weber] entwickelten Gesetze der Schwingungen eines elektrschen Atomenpaares [...] werden wahrscheinlich zur analytischen Bestimmungen der Zahl und Lage der Spectral-Linien der chemischen Elemente und ihres Zusammenhanges mit den Atomgewichten der letzteren führen.

144 [Web71, p. 278 of Weber's *Werke*] and [Web72, p. 129].

145 [Hec96].

1.10 Weber's Mature Planetary Model of the Atom and the Periodic System of the Elements

Weber's mature planetary model of the atom was presented in his work *Determinations of electrodynamic measure: particularly in respect to the connection of the fundamental laws of electricity with the law of gravitation*. This eighth major Memoir, thought to be written in the 1880s, was published posthumously in 1894.[146] It is fragmentary and was not completed in Weber's life time.

In this work Weber presented once more his basic hypothesis of trying to conceive all matter as composed of only two basic building blocks, namely, oppositely charged elementary particles, which he called *electrical molecules* or *simple electrical particles*.[147] The positive elementary particle was represented with an electrical charge $+e$ and an inertial mass ε, while the negative elementary charge was represented with an electrical charge $e' = -e$ and an inertial mass $\varepsilon' = a\varepsilon$. Once more it should be remarked that Weber assumed the electrical charges of these two elementary particles to have the same magnitude, $|e'| = e$, but allowed different magnitudes of inertial masses. In his words:[148]

> [...] it does indeed follow, that there is an equality of mass of all positive electrical molecules among themselves, as well as of all negative electrical molecules among themselves, but it by no means follows, that the masses of positive and negative molecules are the same, rather the decision about the equality or inequality of their masses must remain for experiment to determine, be it by direct measurements of mass, or by an indirect route by investigating their connection with other measurable phenomena.

1.10.1 Deriving a Gravitational Force Law from Weber's Electric Force Law

One of Weber's main goals in this work was to connect gravitation with electricity. In order to derive the law of gravitation from his fundamental force law, Weber presented two main assumptions, namely:[149]

1. *That all ponderable molecules are mere connections of equal quantities of positive and negative electricity*, and that

---

146 [Web94a] and [webb].
147 [Web94a, p. 479 and 492] and [webb, p. 1 and 18].
148 [Web94a, p. 482] and [webb, p. 5].
149 [Web94a, p. 481] and [webb, p. 4].

2. *The force of attraction of equal quantities of different kinds of electricity is greater than the repulsive force of the same quantities of similarly charged electricity.*

Hypotheses analogous to these ones had been given before by Mossotti in 1836.[150] He had applied them to the repulsive forces between the molecules of matter, to the repulsive forces between the atoms of the ether, and to the attractive forces between a molecule of matter and an atom of the ether. In this way he derived a force analogous to Newton's law of gravitation. Zöllner, on the other hand, applied these hypotheses qualitatively for the electrostatic potentials.[151] That is, he assumed that the attractive electrostatic potential between two oppositely charged particles would be a little larger than the repulsive electrostatic potential between two positively charged particles, or between two negatively charged particles. In this way he obtained an attractive force between two binary assemblies (with each binary assembly being composed of two oppositely charged particles $+e$ and $-e$) similar to Newton's law of gravitation.

Weber considered a similar idea, but applied it quantitatively to his fundamental force law. In this way he derived a force law analogous to Newton's law for gravitation, but now including terms which depended upon the relative velocity and relative acceleration between the interacting masses.

According to the first hypothesis, each ponderable molecule would be composed of an integer number n of positive elementary positive particles, and an equal number n of negative elementary negative particles. That is, it would be electrically neutral, being composed of $+ne$ positive elementary charges and $-ne$ negative elementary charges.

Let us now illustrate the utilization of the second hypothesis in order to derive an analogous to Newton's law of gravitation. The simplest ponderable molecule would be composed of a positive elementary charge and a negative elementary charge orbiting around one another. As was seen in Section 1.6.5, Weber had obtained in 1871 a solution for the problem of two charges of opposite sign interacting with one another according to his force law. This particular solution was analogous to the Keplerian elliptical orbit obtained from Newton's law of gravitation. Let us now consider the interaction of two simple molecules according to these two hypotheses. In order to clarify Weber's reasoning, we represent here the two oppositely charged particles of the first molecule by $+e_1$ and $-e_1$, while the two oppositely charged particles of the second molecule are represented by $+e_2$ and $-e_2$, respectively. The indexes 1 and 2 are introduced

---

150 [Mos36].
151 [Zol78a] and [Zol82].

here only to distinguish the two molecules, as $|e_1| = |e_2| = e$. The interaction between these two simplest ponderable molecules would be composed of four forces, namely, the force between $+e_1$ and $+e_2$, that between $+e_1$ and $-e_2$, that between $-e_1$ and $+e_2$, and, finally, that between $-e_1$ and $-e_2$. The sum of these four interactions results in a net attraction between the first ponderable molecule and the second ponderable molecule, due to the second hypothesis.

1.10.2 The Manifold of Ponderable Bodies

According to Weber, the positive and negative particles composing the ponderable molecules should not be considered as concentrated in a single point. That is, they should be considered as always in motion and separated from one another. They could orbit around one another, or could vibrate along the straight line connecting them or orthogonally to this direction:[152]

> But even if the two equal amounts of two dissimilar electrical molecules $+e$ and $-e$ can combine into a ponderable molecule, there will be no combination into a point, rather, as close to one another as the two molecules may come, they will still always remain separated from one another, in that they rotate around one another; the two, which together have the mass of $(1 + a)\varepsilon$, will, however, always remain in a very small space [of volume v], which does not change under conditions of unchanged angular velocity, so that a certain density $d = [(1 + a)/v] \cdot \varepsilon$ can be ascribed to such a ponderable molecule.

Although Weber wanted to reduce all ponderable matter as a combination of only two types of elementary charged particles, he knew there were a great number of ponderable substances with a variety of intrinsic properties. How to explain this enormous complexity out of such simple elements? Here is Weber's nswer:[153]

> But were all ponderable bodies really only combinations of positive and negative electrical molecules, the issue would be, given the essentially identical constitution of all ponderable bodies, how to explain the infinite multiplicity and difference of these ponderable bodies. The reason for all of these differentiations could only be found in different numbers, spatial arrangements and kinetic energy of the

---

[152] [Web94a, p. 490] and [webb, p. 16].
[153] [Web94a, p. 491] and [webb, p. 17].

electrical molecules of both type combined in smaller groups, which need not be subjected to changes by external influences.

According to Weber, some of these arrangements would be extremely stable and could not be changed through external influences. But other groups would not be so stable and could be modified by appropriate external interactions.

As seen in Chapter 1.6, according to Weber's law it is possible to have an attraction of two or more charged particles of the same sign if they are very close to one another. In particular, if the distance r between any two charges of this group is initially smaller than the critical distance ϱ, they would always remain moving relative to one another inside a sphere of diameter ϱ. In 1871 Weber coined the name *molecular motion* to this state of aggregation of charged particles of the same sign always attracting one another while in relative motion. He now gave another name to this group, namely, *indissoluble molecules* [*unscheidbarer Moleküle*].[154] He characterized them in a beautiful way by saying that this group formed an enclosed world for itself, due to the fact that the internal force connecting the group would be so great that it would be extremely difficult to break it apart due to external influences. Here is a typical quotation of Weber:[155]

> *Additionally, not only two or three, but a far larger number of similar electrical particles could be together in such a small space, without the distance of any particle from another being greater than or equal to ϱ, so that all of these particles together, also form an indissoluble molecule which remains together for ever. And finally, it should be noted that these particles enclosed in a small space of a molecule, have as little need to be at rest as the particles originally dispersed in larger spaces, but they can have the most manifold movements, partly together, in close connection with one another in space, partly against one another within the small space in which they are, without thereby ceasing to form an indissoluble group or a single composite molecule. Each such composite molecule forms an enclosed world for itself, and according to the difference of the number of simple electrical particles which it contains, and their mutual movements, such a composite molecule can exert quite diverse effects upon all other molecules lying outside of it, according to which very diverse characteristics may obtain for that molecule. If one further considers, that the number of simple electrical particles which can be combined in this way, although not unlimited, can*

---
154 [Web94a, p. 492–493] and [webb, p. 18–19].
155 [Web94a, p. 493] and [webb, p. 19].

> *yet be very large, it is conceivable, that such eternally unchangeable, partly positive, partly negative electrical particles or molecules can recombine themselves to quite different ponderable bodies, for example of very different density or hardness, etc., for each group consisting of a larger number of similar electrical particles, partly positive, partly negative, of which each occupies only a spherical space of diameter ϱ, must obviously attract each other and combine with a force much larger than a simple positive electrical molecule with a simple negative electrical molecule.*

Weber could then classify his material molecules into three separate classes.

1. The first class was composed of his two elementary charges $+e$ and $e' = -e$ having inertial masses ε and $\varepsilon' = a\varepsilon$, respectively. He called them *simple positive and negative electrical molecules* and represented them by the symbols $(+1)$ and (-1).

2. The second class included his positive and negative indissoluble electrical molecules. Each of these composite molecules would have n elementary charges of the same sign interacting with one another inside a very small sphere of diameter ϱ, with $n \geq 2$. They were represented by the symbol $(+n)$ or $(-n)$, depending upon the sign of the constituent participants. For instance, we could have a composite indissoluble positive electrical molecule composed of three positive charges, $(+3)$, or a composite indissoluble negative electrical molecule composed of five negative charges, (-5).

3. The third class contained the ponderable molecules, which were arrangements having the same number of positive and negative electrical molecules.

By continuing his analysis, Weber found that each of these ponderable molecules would exert an attractive force not only upon another ponderable molecule, but also upon an elementary positive charge and upon an elementary negative charge. By means of this force of attraction exerted by a ponderable molecule on the positive (or negative) electrical elementary charge, the latter could be continuously maintained in a rotational motion around the ponderable molecule. In this way he could imagine the existence of positive and negative ponderable ions, although he did not utilize this modern expression "ion".

This allowed Weber to classify the ponderable molecules into three separated groups, namely, neutral ponderable molecules, positive ponderable ions and negative ponderable ions. He expressed this as follows:[156]

> By virtue of the here postulated force of attraction exerted by every ponderable molecule, not merely upon another equal molecule, but upon each of its two constituent parts, all of those ponderable molecules which had first met up with positive electrical molecules, would have bound them as positive electrical satellites, and, on the other hand, other entirely identical ponderable molecules, which had first met up with negative electrical molecules, would have bound them as negative electrical satellites; and, therefore, all ponderable molecules would fall into three classes, which can be distinguished as positive ponderable, negative ponderable, and neutral, of which the latter would be such ponderable molecules, which had not yet drawn satellites to themselves.

1.10.3 The Periodic Table of the Chemical Elements

According to Weber, the neutral ponderable molecules would be composed of the same number of positive and negative electrical molecules.

He represented them graphically as in Figure 1.7:[157]

Here we clarify the meaning of one of these neutral ponderable molecules by choosing a specific example, namely, the molecule $\begin{bmatrix} +3 \\ -2 \\ -1 \end{bmatrix}$. The number $+3$ would represent three elementary charges $+e$ moving about each other in such a way that the distance of any two of these charges would be always smaller than the critical distance ϱ. This inseparable group, composed of three like charges, would move as a single particle. Likewise, the number -2 would represent another inseparable portion of the ponderable molecule. This group -2 would be composed of two negative elementary charges $-e$ moving about each other in such a way that their distance would be always smaller than the critical distance ϱ. This inseparable group -2 would also move as a single particle. And finally the number -1 would represent a simple negative electrical molecule $-e$. We could then think of this ponderable molecule as being composed of three particles of charges $+3e$, $-2e$ and $-e$, respectively, moving about each other in three different orbits. Weber expressed this as follows:[158]

---

156 [Web94a, p. 485] and [webb, p. 11].
157 [Web94a, p. 495] and [webb, p. 22].
158 [Web94a, p. 494–496] and [webb, p. 23].

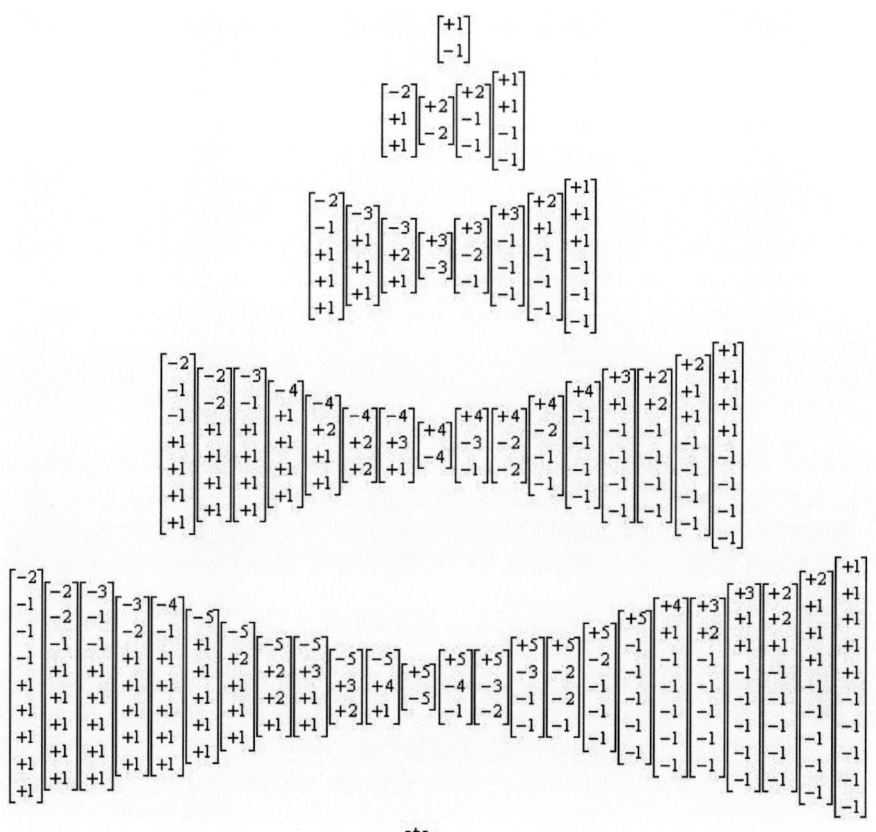

Figure 1.7:
The ponderable molecules according to Wilhelm Weber
[Web94a, p. 495] and [webb, p. 22].

Here, each of the numbers enclosed in the same brackets refers to a number of like electrical particles moving about each other, whose distance from one another remains smaller than ϱ. These indissoluble particles move together in one orbit, and a particular orbit corresponds to each number. The orbits of dissimilar electrical particles are held together by mutual attraction. The molecules comprised in each number are, accordingly, indissoluble and likewise also all of the molecules of the second of the three classes enumerated above, also those, which are composites of many simple molecules, because they are similar and their distances from one another are $< \varrho$.

Weber called the molecules $\begin{bmatrix} +n \\ -n \end{bmatrix}$ as *ponderable elementary bodies*. This might represent either the specific element with two indissoluble electrical particles orbiting around one another (with each particle being composed on n elementary charges moving around one another inside a sphere of diameter ϱ), or the whole set of possibilities represented by the n-th line of Figure 1.7.

With the ponderable elementary bodies $\begin{bmatrix} +n \\ -n \end{bmatrix}$ Weber wanted to characterize the atoms of the chemical elements of the periodic table. The simplest ponderable elementary body is hydrogen with atomic weight $= 1$. It was represented as $\begin{bmatrix} +1 \\ -1 \end{bmatrix}$. In the diagram with five lines of Fig. 1.7, Weber listed all possible configurations of $+e$ and $-e$ from atomic weight $= 1$ up to atomic weight $= 5$. At the n-th line we find the possible configurations of $\begin{bmatrix} +n \\ -n \end{bmatrix}$. All configurations of a single line have the same atomic or molecular weight. A chemical atom with atomic weight n would have the configuration $\begin{bmatrix} +n \\ -n \end{bmatrix}$. Carbon and oxygen, for instance, were represented by Weber as $\begin{bmatrix} +12 \\ -12 \end{bmatrix}$ and $\begin{bmatrix} +16 \\ -16 \end{bmatrix}$, respectively. They would be along the 12th and 16th lines, respectively.

A ponderable molecule would be attracted by another ponderable molecule, so that there could exist *composites of ponderable molecules* orbiting around one another. Weber called these systems by the name *binary composite ponderable bodies*. These would correspond to the *molecules* of modern chemical terminology, that is, compositions of elementary atoms giving rise to stable systems. Weber concluded his analysis of this Section as follows:[159]

> *If one further considers the extraordinary multiplicity which can occur in each of these ponderable molecules in relationship to the orbits and the vis vivas [that is, kinetic energies] of particular electrical particles, out of which they are composed, there is the possibility of infinitely many different kinds of such molecules.*

Between the multiple variations of Weber's *ponderable molecules* (our modern *elementary atoms*) there are, in particular, the following types:

$$\begin{bmatrix} +n \\ -1 \\ \cdot \\ \cdot \\ \cdot \\ -1 \end{bmatrix}$$

Here we have n positive elementary charged particles moving relative to one another in a very small volume of diameter ϱ (the critical distance). The attractive force between these positive particles is so strong, that this positive system behaves itself as a single particle. It would correspond to the nucleus of a modern atom. Around this positive nucleus orbited n negative elementary particles in separate orbits. They would correspond to the negative electrons in orbit around a positive nucleus. As Weber did not equate the inertial mass ε of each elementary positive particle with the inertial mass $\varepsilon' = a\varepsilon$ of each elementary negative particle, the mass of this positive nucleus was not necessarily n times the mass of each negative particle orbiting around it.

It is amazing the analogy of this Weberian conception with the Rutherford-Bohr atomic model.

Weber *indissoluble molecule* (that is, the positive nucleus of this ponderable molecule) would represent the modern nuclei of the atomic elements. However, in Weber's model there is no particle corresponding to the neutron. On the other hand, it has the amazing advantage of being stable and held together by purely electric interactions, without the necessity of separate forces like the weak and strong nuclear forces of modern physics. To our knowledge this is the only model ever proposed of a positive nucleus stabilized by purely electric interactions. Weber could only succeed in obtaining this feature due to a unique

---

159 [Web94a, p. 499] and [webb, p. 27].

property of his force law, Eq. (1.2). This property is related to the fact that his force law depends not only upon the distance between the interacting particles, but also upon their relative radial acceleration. The coefficient multiplying this acceleration has the same unit as that of inertial mass, namely, kg. Moreover, this coefficient is proportional to the product of the two interacting charges, ee', and is inversely proportional to their distance r. When they are very close to one another, this coefficient can have a magnitude greater than the mechanical inertial mass of any of these particles. These charges can then behave as if they had an effective inertial mass which is a function of the distance separating them. Moreover, this effective inertial mass can be positive or negative, depending upon the sign of ee'. In particular, charges of the same sign moving relative to one another inside a sphere of diameter ϱ will behave as having an effective *negative* inertial mass. Consequently, instead of repelling one another as usually observed at macroscopic distances, they will attract one another! This is one of the most fascinating and unique properties of Weber's electrodynamics.[160]

1.10.4 Application to Chemical Bondings

As was seen in Chapter 1.6, Weber had analyzed in 1871 the motion of two charged particles interacting according to his force law. He found some *permanent states* in which the particles would orbit around one another between two limiting radii. He also found some states with *want of permanence* in which the two interacting particles would not be connected to one another. This would be analogous to the Coulombian hyperbolic orbits of two charges of opposite sign attracting one another.[161] In that work of 1871 Weber made a bold hypothesis, namely, that the chemical atomic bonds between atoms might have an electrical origin. In chemistry we have some pairs of atoms, as the gases H_2 or O_2, and also several molecules composed of atoms, like H_2O etc. If Weber's hypothesis were really true, his electrical states of aggregation might have an applicability to chemical atomic groups. He expressed himself as follows:[162]

Applicability to Chemical Atomic Groups.

> [...] *The question consequently presents itself, whether the cause of the permanence of certain atomic states may not perhaps be found*

---

[160] [23], [1] and [2].
[161] [Sym71, Section 3.16].
[162] [Web71, p. 278–279 of Weber's *Werke*] and [Web72, p. 129–130].

in such law of mutual action as have here been established and assumed for electrical particles. [...] And in relation to this it is to be specially observed that the same forces as those which determine the states of aggregation of electricity formed by simple atoms and by atomic pairs, may possibly also determine similar states of aggregation of ponderable bodies. For in the general distribution of electricity it must be assumed that an atom of electricity adheres to each ponderable atom. [...] If now we take the values of [the masses] ε and ε' so great as to include the masses of the ponderable atoms adhering to the electrical atoms, all the results that have been arrived at in reference first of all to electrical atoms merely, may also be applied to the ponderable atoms combined with the electrical atoms.

That is, assuming the possibility that a positive or a negative electrical particle might adhere to a ponderable atom, there would exist permanent states of aggregation of chemical atoms and also states of aggregation with want of permanence. Weber's electrodynamic force law might then offer a key in order to understand the chemical bonds of atoms.

In his posthumous work Weber developed a little more this idea. He presented here a model which might have a relation with an electrochemical bond. He wanted to understand the solid cohesion of metals. As seen in Chapter 1.5, since 1852 Weber had a model of a metallic conductor like copper. He considered this conductor as a sequence of positive ponderable atoms separated in space, with each atom being surrounded by an orbiting negative particle of negligible mass compared to the mass of the ponderable atom. Later on he changed the signs of the stationary and orbiting particles. Until the 1880's the sign of the orbiting charge had not yet been found out experimentally. He now presented a reason for the cohesion of solid substances:[163]

The really very solid cohesion of the ponderable molecules of metallic conductors is probably due to the reason, that each positive electrical molecule in its circular orbit, encompasses not only the one negative electrical molecule of the one ponderable neighboring molecule, but also the other negative electrical molecule of the other ponderable neighboring molecule. The same holds for the circular orbit of every negative electrical molecule and two positive electrical neighboring molecules.

---

[163] [Web94a, p. 508] and [webb, p. 27].

According to our present knowledge, this does not correspond to the correct metallic bond. In any event, Weber was here presenting a new idea. Humphry Davy (1778–1829) and Jöns Jacob Berzelius (1779–1848) had already proposed in 1806 and 1811 (and 1819), respectively, the utilization of Coulomb's force in the electrochemical theories.[164] Weber went a step further in presenting a *dynamical* electrical force in order to explain chemical bonds. This model presented by Weber in his posthumous work is analogous to the covalent bond developed by Gilbert Newton Lewis (1875–1946) in 1916, which is based on a bond through an electron pair being shared between two neighboring atoms.

1.10.5 Open Topics

In the sequence of his posthumous paper, Weber pointed out some problems which still needed to be solved by following the hypothesis that the ponderable molecules were composed of positive and negative elementary charged particles. For instance, it would be necessary to explain the mechanics of expandable fluids (gases), of non-expandable fluids (liquids) and of solid elastic bodies. He did not pretend to solve these extremely complex problems out of his fundamental law of electric force. He only presented a few suggestions. His ideas here, as in his paper of 1871 related to molecular movements, were characterized as follows:[165]

> *For so long as the molecular forces acting only at molecular distances, which doubtless cooperate in the molecular movements, are not known and taken exact account of, the results that may be acquired cannot have any exact quantitative application, but only a qualitative value within certain limits, and can be of consequence only for a first reconnaissance of the territory.*

In order to explain the expansion of gases, for instance, Weber thought that its molecules might consist of ponderable nuclei moving at large distances to one another, each accompanied by a positive electrical satellite. The molecules of non-expandable fluids, on the other hand, were supposed to "consist of *ponderable molecules without satellites* which, on account of the reciprocal *force of gravitation* exerted upon each other, would rotate around each other." As regards conducting solids, he proposed once more a

> molecular constitution, according to which positive electrical molecules rotate in these bodies around the individual ponderable mo-

---

164 [Wie60, p. 219–220] and [Wie67, p. 177].
165 [Web71, p. 269 of Weber's *Werke*] and [Web72, p. 119].

> *lecules with continuously changing radii, each for so long, until it is transposed into a ballistic motion, and is thereby led out of the sphere of action of one ponderable molecule into that of another.*

And finally for insulating solids, he followed Mossotti's idea,

> *according to which molecules at certain distances from each other are in stable equilibrium, which equilibrium comes into being through the repulsive forces of these molecules themselves, also through the repulsive forces of molecules of a (positive electrical) fluid contained in the intervening spaces, and finally through the forces of attraction between the ponderable and the (positive electrical) molecules.*

Weber presented some general ideas which might help to explain the differences between ice, water and steam out of his fundamental force law. He also discussed in general terms the melting point of ice, the boiling point of water and the crystal formation of solid bodies.

1.11 Final Considerations

In this final Chapter we call attention to some speculations and developments of Weber's atomistic conception of nature and his planetary model of the atom.

Cathode rays were studied by many scientists at the end of the XIXth century. During many years it was not clear if they were a wave phenomena or if they were due to mobile charged particles. Heinrich Hertz (1857–1894), for instance, made experiments related to this subject. He was strongly influenced by Helmholtz (1821–1894), an opponent of Weber. Helmholtz and Hertz believed that the cathode rays were a wave phenomena. However, later on it was shown conclusively that cathode rays were due to the motion of charged particles.

In 1881 Eduard Riecke (1845–1915), successor of Wilhelm Weber in Göttingen and advocate of Weber's atomistic conceptions, utilized in his publication of 1881 Weber's atoms and the corresponding symbols.[166] For instance, the symbol e was utilized for the electric charge of the particle and the symbol ε was utilized for the inertial mass of the particle. With this procedure, Riecke succeeded in deducing the circular and spiral orbits of a charged particles moving in an homogeneous magnetic field.

---

166 [Rie81] and [Wie08].

In 1878 Zöllner (cf. Figure 3.2, p. 162) published a paper discussing the electric phenomena produced by light and radiating heat.[167] In particular, he discussed the discovery of Hankel in 1877 of the production of electrostatic charges in the surface of crystals through light radiation and the discovery of Sale in 1873 that the crystal selenium changes its resistance according to the frequency of light incident upon it. In the conclusion of his paper Zöllner made the following remark:[168]

> *According to the electrodynamic theory of matter and to W. Weber's point of view this means that the electric particles, which are connected to one another in molecular currents, can be separated from one another and disposed in other orbits of their motion through irradiation.*

This is an amazing insight coming many years before Hertz discovery of the photoelectric effect in 1887. Weber considered that the ponderable atoms and molecules were held together through electrodynamic forces acting between its elementary charged particles. This opened the possibility of considering the transmutation of one chemical element into another through external electrodynamic forces![169] Weber, in particular, made the following comments into his last major Memoir:[170]

> *In all molecules of the third class, on the other hand, those listed as positive electrical under + are possibly always dissoluble from the negative electrical listed under −, even if no force sufficient to cause their dissolution exists. In reality, no such dissolution, whereby a ponderable body were broken up into its imponderable constituent parts, has been observed. But since the dissolution of ponderable bodies into ponderable constituent parts is often observed, by continued dissolution, however, one finally arrives at ponderable bodies which have not been further dissoluble, one has indeed called these latter ponderable bodies elemental bodies, whereby however the possibility of their dissolution into positive and negative electrical molecules is not excluded.*

---

167 [Zol78b].
168 [Zol78b, p. 610]:
   Vom Standpunkte der elektrodynamischen Theorie der Materie und im Sinne der Anschauungen W. Weber's heisst dies aber nichts Anderes, als dass elektrische Theilchen, welche zu Molecularströmen mit einander vereinigt sind, durch Bestrahlung von einander getrennt und zu andern Bahnen ihrer Bewegung veranlasst werden können.
169 [Wie60, p. 216–220].
170 [Web94a, p. 496] and [webb, p. 23].

Friedrich Kohlrausch (1840–1910) (cf. Figure 2.6, S. 115) was one of the physicists who perceived the possibility of transmutation of the chemical elements through electrodynamic forces. He was the son of Weber's collaborator, Rudolf Kohlrausch, and was initially a salaried assistant of Weber at the Physics Institute of Göttingen University (1866). In 1870 Weber resigned the directorship of Göttingen's seminar so that F. Kohlrausch could take this position.[171] Weber was also his doctoral advisor. In 1879, when discussing Weber's planetary model of the atom, F. Kohlrausch expressed a genial idea, namely:[172]

> *If we admit here the possibility, that the electricity and this primordial matter of the substance are identical or are closely associated with one another, then perhaps electrical means would offer the best chance of success in transmutating chemically different substances into one another.*

This possibility has been realized in practice half a century later in the particle accelerators of modern physics.

1.12 Acknowledgments

One of the authors, AKTA, wishes to thank the Institute for the History of Natural Sciences of Hamburg University and the Alexander von Humboldt Foundation of Germany for a research fellowship in 2009 during which this work was accomplished. He was invited by Prof. Dr. G. Wolfschmidt and continued the research he had performed at Hamburg University in 2001–2002, invited by Prof. K. Reich, with an earlier fellowship also granted by the Humboldt Foundation. He thanks also Faepex/UNICAMP for financial support and the Institute of Physics of UNICAMP for supplying the necessary conditions in order to undertake this work.

The authors thank a number of people for support, suggestions, corrections, references etc.: B. R. Bligh, C. Blondel, J. Gottschalk, L. Hecht, W. Lange, T. E. Phipps Jr., K. Reich, T. Rüting, J. Tennenbaum, B. Wolfram and G. Wolfschmidt.

---

171 [JM86, p. 104–107].
172 [Koh11, Vol. 2, p. 172] and [Wie60, p. 218–219]:
   Will man hiernach die Möglichkeit zugeben, daß die Elektrizität und dieser Urstoff der Substanz identisch sind oder doch in nahem Zusammenhang stehen, so würden bei dem Versuche, chemisch verschiedene Stoffe ineinander zu verwandeln, elektrische Hilfsmittel vielleicht die größte Aussicht auf einen Erfolg bieten.

Bibliography

[Amp20a] AMPÈRE, A.-M.: Mémoire présenté à l'Académie royale des Sciences, le 2 octobre 1820, où se trouve compris le résumé de ce qui avait été lu à la même Académie les 18 et 25 septembre 1820, sur les effets des courans électriques. In: *Annales de Chimie et de Physique* 15 (1820), p. 59–76.

[Amp20b] AMPÈRE, A.-M.: Suite du Mémoire sur l'Action mutuelle entre deux courans électriques, entre un courant électrique et un aimant ou le globe terrestre, et entre deux aimants. In: *Annales de Chimie et de Physique* 15 (1820), p. 170–218.

[Amp22a] AMPÈRE, A.-M.: Expériences relatives à de nouveaux phénomènes électro-dynamiques. In: *Annales de Chimie et de Physique* 20 (1822), p. 60–74.

[Amp22b] AMPÈRE, A.-M.: Mémoire sur la Détermination de la formule qui représente l'action mutuelle de deux portions infiniment petites de conducteurs voltaïques. Lu à l'Académie royale des Sciences, dans la séance du 10 juin 1822. In: *Annales de Chimie et de Physique* 20 (1822), p. 398–421.

[Amp23] AMPÈRE, A.-M.: Mémoire sur la théorie mathématique des phénomènes électro-dynamiques uniquement déduite de l'expérience, dans lequel se trouvent réunis les Mémoires que M. Ampère a communiqués à l'Académie royale des Sciences, dans les séances des 4 et 26 décembre 1820, 10 juin 1822, 22 décembre 1823, 12 septembre et 21 novembre 1825. In: *Mémoires de l'Académie Royale des Sciences de l'Institut de France* 6 (1823), p. 175–387. Despite the date, this work was only published in 1827.

[Amp26] AMPÈRE, A.-M.: *Théorie des Phénomènes Électro-dynamiques, Uniquement Déduite de l'Expérience.* Paris: Méquignon-Marvis 1826.

[Amp65a] AMPÈRE, A.-M.: The effects of electric currents. In: TRICKER, R. A. R.: *Early Electrodynamics – The First Law of Circulation.* Translated by O. M. BLUNN. New York: Pergamon 1965, p. 140–154.

[Amp65b] AMPÈRE, A.-M.: On the Mathematical Theory of Electrodynamic Phenomena, Experimentally Deduced. In: TRICKER, R. A. R.: *Early Electrodynamics – The First Law of Circulation.* New York: Pergamon 1965, p. 155–200. Partial translation by O. M. BLUNN of Ampère's work *Mémoire sur la théorie mathématique des phénomènes électro-dynamiques uniquement déduite de l'expérience.*

[AC92] ASSIS, A. K. T. AND R. A. CLEMENTE: The ultimate speed implied by theories of Weber's type. In: *International Journal of Theoretical Physics* 31 (1992), p. 1063–1073.

[AH07] ASSIS, A. K. T. AND J. A. HERNANDES: *The Electric Force of a Current: Weber and the Surface Charges of Resistive Conductors Carrying Steady Currents*. Montreal: Apeiron 2007. ISBN: 978-0-9732911-5-5. Available at: http://www.ifi.unicamp.br/~assis.

[ARW02] ASSIS, A. K. T.; REICH, K. AND K. H. WIEDERKEHR: Gauss and Weber's creation of the absolute system of units in physics. In: *21st Century* 15 (2002), No. 3, p. 40–48.

[ARW04] ASSIS, A. K. T.; REICH, K. AND K. H. WIEDERKEHR: On the electromagnetic and electrostatic units of current and the meaning of the absolute system of units – For the 200th anniversary of Wilhelm Weber's birth. In: *Sudhoffs Archiv* 88 (2004), p. 10–31.

[Ass89] ASSIS, ANDRE KOCH TORRES: On Mach's principle. In: *Foundations of Physics Letters* 2 (1989), p. 301–318.

[Ass93] ASSIS, ANDRE KOCH TORRES: Changing the inertial mass of a charged particle. In: *Journal of the Physical Society of Japan* 62 (1993), p. 1418–1422.

[Ass94] ASSIS, ANDRE KOCH TORRES: *Weber's Electrodynamics*. Dordrecht: Kluwer Academic Publishers 1994. ISBN: 0-7923-3137-0.

[Ass97] ASSIS, ANDRE KOCH TORRES: Circuit theory in Weber electrodynamics. In: *European Journal of Physics* 18 (1997), p. 241–246.

[Ass99] ASSIS, ANDRE KOCH TORRES: *Relational Mechanics*. Montreal: Apeiron 1999. ISBN: 0-9683689-2-1. Available at: http://www.ifi.unicamp.br/~assis.

[AW03] ASSIS, A. K. T. AND K. H. WIEDERKEHR: Weber quoting Maxwell. In: *Mitteilungen der Gauss-Gesellschaft* 40 (2003), p. 53–74.

[Bio28] BIOT, J. B.: *Lehrbuch der Experimentalphsik oder Erfahrungsnaturlehre*, vol. 1. Leipzig: Leopold Voss (2nd edition) 1828. German translation by G. T. FECHNER.

[Blo82] BLONDEL, C.: *A.-M. Ampère et la Création de l'Électrodynamique (1820–1827)*. Paris: Bibliothèque Nationale 1982.

[CA91] CLEMENTE, R. A. AND ANDRE KOCH TORRES ASSIS: Two-body problem for Weber-like interactions. In: *International Journal of Theoretical Physics* 30 (1991), p. 537–545.

[Dar00] DARRIGOL, O.: *Electrodynamics from Ampère to Einstein*. Oxford: Oxford University Press 2000.

[Dru05] DRUDE, P.: Electric conduction in metals, from the standpoint of the electronic theory. In: *Transactions of the International Electrical Congress*

(St. Louis) 1 (1905), p. 317–330. Reprinted in DRUDE, P.: *Zur Elektronentheorie der Metalle*. Ed. by GRAHN, H. T. AND D. HOFFMANN. Frankfurt: Verlag Harri Deutsch (Ostwalds Klassiker der exakten Wissenschaften; Vol. 298) 2006.

[Eby77] EBY, P. B.: On the perihelion precession as a Machian effect. In: *Lettere al Nuovo Cimento* 18 (1977), p. 93–96.

[Far21a] FARADAY, MICHAEL: Historical sketch of electro-magnetism. In: *Annals of Philosophy* 2 (1821), p. 195–200.

[Far21b] FARADAY, MICHAEL: Historical sketch of electro-magnetism. In: *Annals of Philosophy* 2 (1821), p. 274–290.

[Far22] FARADAY, MICHAEL: Historical sketch of electro-magnetism. In: *Annals of Philosophy* 3 (1822), p. 107–121.

[Far65a] FARADAY, MICHAEL: *Experimental Researches in Electricity*, vol. I and II. New York: Dover 1965.

[Far65b] FARADAY, MICHAEL: *Experimental Researches in Electricity*, vol. III. New York: Dover 1965.

[fec] FECHNER, G. T.: Drei Versuchsreihen, welche die elektromotorische Kraft und die einzelnen Elemente des Leitungswiderstandes betreffen. In: BIOT, J. B.: *Lehrbuch des Galvanismus und der Elektrochemie*. Leipzig 1829 (2nd edition), Nachträge, p. 553–564. Reprint: Ed. by C. PIEL. Leipzig: Akademische Verlagsgesellschaft (Ostwald's Klassiker der exakten Wissenschaften; Nr. 244) 1938, p. 30–38.

[Fec45] FECHNER, G. T.: Ueber die Verknüpfung der Faraday'schen Inductions-Erscheinungen mit den Ampèreschen elektro-dynamischen Erscheinungen. In: *Annalen der Physik* 64 (1845), p. 337–345.

[Fec64] FECHNER, G. T.: *Ueber die physikalische und philosophische Atomenlehre*. Leipzig: Hermann Mendelssohn (2nd edition) 1864.

[Fre85a] FRESNEL, A.: Comparaison de la supposition des courants autour de l'axe avec celle des courants autour chaque molécule. In: Joubert, J. (ed.): *Collection de Mémoires relatifs a la Physique, Vol. II: Mémoires sur l'Électrodynamique*. Paris: Gauthier-Villars 1885, p. 141–143.

[Fre85b] FRESNEL, A.: Deuxième note sur l'hipothèse des courants particulaires. In: JOUBERT, J. (ed.): *Collection de Mémoires relatifs a la Physique, Vol. II: Mémoires sur l'Électrodynamique*. Paris: Gauthier-Villars 1885, p. 144–147.

[Ger98] GERBER, P.: Die räumliche und zeitliche Ausbreitung der Gravitation. In: *Zeitschrift fur Mathematik und Physik II* 43 (1898), p. 93–104.

[Ger17] GERBER, P.: Die Fortpflanzungsgeschwindigkeit der Gravitation. In: *Annalen der Physik* 52 (1917), p. 415–444.

[GG90] GRATTAN-GUINNESS, I.: *Convolutions in French Mathematics, 1800–1840*, vol. 2. Basel: Birkhäuser 1990.

[GG91] GRATTAN-GUINNESS, I.: Lines of mathematical thought in the electrodynamics of Ampère. In: *Physis* 28 (1991), p. 115–129.

[Hec96] HECHT, L.: The significance of the 1845 Gauss-Weber correspondence. In: *21st Century* 9(3) (1996), p. 22–34.

[Hei93] HEIDELBERGER, M.: *Die innere Seite der Natur: Gustav Theodor Fechner wissenschaftlich-philosophische Weltauffassung*. Frankfurt: Vittorio Klostermann 1993.

[Hei04] HEIDELBERGER, M.: *Nature from Within: Gustav Theodor Fechner and His Psychophysical Worldview*. Translated by C. KLOHR. Pittsburgh: University of Pittsburgh Press 2004.

[Hof82] HOFMANN, J. R.: *The Great Turning Point in André-Marie Ampère's Research in Electrodynamics: A Truly "Crucial" Experiment*. PhD thesis, Graduate Faculty of Arts and Sciences. Pittsburgh: University of Pittsburgh 1982.

[Hof87] HOFMANN, J. R.: Ampère, electrodynamics, and experimental evidence. In: *Osiris* 3 (2nd Series) (1987), p. 45–76.

[Hof96] HOFMANN, J. R.: *André-Marie Ampère, Enlightenment and Electrodynamics*. Cambridge: Cambridge University Press 1996.

[JM86] JUNGNICKEL, C. AND R. MCCORMMACH: *Intellectual Mastery of Nature — Theoretical Physics from Ohm to Einstein*, volume 1–2. Chicago: University of Chicago Press 1986.

[Koh11] KOHLRAUSCH, F.: *Gesammelte Abhandlungen von Friedrich Kohlrausch*. 2 volumes. Ed. by W. HALLWACHS, A. HEYDWEILLER, K. STRECKER AND O. WIENER. Leipzig: J. A. Barth 1910–1911.

[KW57] KOHLRAUSCH, R. AND W. WEBER: Elektrodynamische Maassbestimmungen insbesondere Zurückführung der Stromintensitäts-Messungen auf mechanisches Maass. In: *Abhandlungen der Königl. Sächs. Gesellschaft der Wissenschaften, mathematisch-physische Klasse* 3 (1857), p. 221–290.
Reprinted in Wilhelm Weber's *Werke*, Vol. 3, ed. by H. WEBER. Berlin: Springer 1893, p. 609–676.

[Max83] MAXWELL, J. C.: *Lehrbuch der Electricität und des Magnetismus*. 2 Bde. Deutsche Übersetzung von B. WEINSTEIN. Berlin: Springer 1883.

[Max54a] MAXWELL, J. C.: *A Treatise on Electricity and Magnetism, vol. I*. New York: Dover 1954.

[Max54b] MAXWELL, J. C.: *A Treatise on Electricity and Magnetism, vol. II*. New York: Dover 1954.

[Mos36] MOSSOTTI, O. F.: *Sur les forces qui régissent la constitution intérieure des corps, apperçu pour servir à la détermination de la cause et des lois de l'action moléculaire*. Turin: L'Imprimerie Royale 1836. Reprinted in ZÖLLNER, F.: *Erklärung der Universellen Gravitation aus den statischen Wirkungen der*

Elektricität und die allgemeine Bedeutung des Weber'schen Gesetzes. Leipzig: L. Staackmann 1882, p. 83–112.

[Neu58] NEUMANN, C.: *Explicare tentatur quomodo fiat ut lucis planum polarisationis per vires electricas vel magneticas declinetur.* Halis Saxonum 1858.

[Neu63] NEUMANN, C.: *Die magnetische Drehung der Polarisationsebene des Lichts.* Halle: Verlag des Buchhandlung des Waisenhauses 1863.

[Nor65] NORTH, J. D.: *The Measure of the Universe – A History of Modern Cosmology.* Oxford: Clarendon Press 1965.

[Oer20] ØRSTED, H. C.: Experiments on the effect of a current of electricity on the magnetic needle. In: *Annals of Philosophy* 16 (1820), p. 273–277. Translated from a printed account drawn up in Latin by the author and transmitted by him to the Editor of the *Annals of Philosophy*.

[Ohm26] OHM, G. S.: Bestimmung des Gesetzes, nach welchem Metalle die Kontakt-Elektrizität leiten, nebst einem Entwurfe zu einer Theorie des Voltaschen Apparates und des Schweiggerschen Multiplikators. In: *Journal für Chemie und Physik* 46 (1826), p. 137–166. Reprint: Ed. by C. PIEL. Leipzig: Akademische Verlagsgesellschaft (Ostwald's Klassiker der exakten Wissenschaften; Nr. 244) 1938, p. 8–29.

[O'R65] O'RAHILLY, A.: *Electromagnetic Theory – A Critical Examination of Fundamentals.* New York: Dover 1965.

[Ørs98] ØRSTED, H. C.: New electro-magnetic experiments. In: JELVED, K.; JACKSON, A. D. AND O. KNUDSEN (ed.): *Selected Scientific Works of Hans Christian Oersted.* Princeton: Princeton University Press 1998, p. 421–424. Paper originally published in German in 1820.

[ram] RAMSAUER, C.: Das Ohmsche Gesetz (1826). In: RAMSAUER, C.: *Grundversuche der Physik in historischer Darstellung.* Vol. 1: Von der Fallgesetzen bis zu den elektrischen Wellen. Berlin: Springer 1953.

[Rie81] RIECKE, E.: über die Bewegung eines elektrischen Teilchens in einem homogenen magnetischen Felde und das negative elektrische Glimmlicht. In: *Annalen der Physik* 13 (1881), p. 191–194.

[Rie92] RIECKE, E.: *Wilhelm Weber (geb. 24. October 1804, gest. 23. Juni 1891).* Rede gehalten in der öffentlichen Sitzung der K. Gesellschaft der Wissenschaften am 5. December 1891. Göttingen: Dieterichsche Verlags-Buchhandlung 1892.

[Sch25] SCHRÖDINGER, E.: Die Erfüllbarkeit der Relativitätsforderung in der klassischen Mechanik. In: *Annalen der Physik* 77 (1925), p. 325–336.

[Sch95] SCHRÖDINGER, E.: The possibility of fulfillment of the relativity requirement in classical mechanics. In: BARBOUR, J. B. AND H. PFISTER (ed.): *Mach's Principle – From Newton's Bucket to Quantum Gravity.* Translated by J. B. BARBOUR. Boston: Birkhäuser 1995, p. 147–158.

[see] SEEGERS, C.: De motu perturbationibusque planetarum secundum legem electrodynamicam Weberianam solem ambientium. Dissertation, Göttingen 1864.

[See24] SEEGERS, C.: *Über die Bewegung und die Störungen der Planeten, wenn dieselben sich nach dem Weberschen elektrodynamischen Gesetz um die Sonne bewegen*. Neu herausgegeben von P. HEYLANDT. Übersetzt von F. Diestel. Braunschweig: Kommissionsverlag von Friedr. Vieweg & Sohn 1924.

[ser] SERVUS, ARMINIUS: Untersuchungen über die Bahn und die Störungen der Himmelskörper mit Zugrundelegung des Weber'schen electrodynamischen Gesetzes. Dissertation, Halle 1885.

[Ste03] STEINLE, F.: The practice of studying practice: Analyzing research records of Ampère and Faraday. In: HOLMES, F. L.; RENN, J. AND H.-J. RHEINBERGER (ed.): *Reworking the Bench: Laboratory Notebooks in the History of Science*. Dordrecht: Kluwer 2003, p. 93–117.

[Ste05] STEINLE, F.: *Explorative Experimente. Ampère, Faraday und die Ursprünge der Elektrodynamik*. Stuttgart: Franz Steiner Verlag 2005.

[Sym71] SYMON, K. R.: *Mechanics*. Reading: München: Addison-Wesley (third edition) 1971.

[Tis72] TISSERAND, FRANÇOIS: Sur le mouvement des planètes autour du soleil, d'après la loi électrodynamique de Weber. In: *Comptes Rendues de l'Academie des Sciences de Paris* 75 (1872), p. 760–763.

[Tis95] TISSERAND, FRANÇOIS: *Traité de Mécanique*, vol. 4, Chapter 28 (Vitesse de propagation de l'attraction) "Loi d'attraction conforme à la loi électrodynamique de Weber". Paris: Gauthier-Villars 1895, p. 499–503.

[TKG23] TOLMAN, R. C.; KARRER, S. AND W. W. GUERNSEY: Further experiments on the mass of the electric carrier in metals. In: *Physical Review* 21 (1923), p. 525–539.

[TMS26] TOLMAN, R. C. AND L. M. MOTT-SMITH: A further study of the inertia of the electric carrier in copper. In: *Physical Review* 28 (1926), p. 794–832.

[TOS14] TOLMAN, R. C.; OSGERBY, E. W. AND T. D. STEWART: The acceleration of electrical conductors. In: *Journal of the American Chemical Society* 36 (1914), p. 466–485.

[TS16] TOLMAN, R. C. AND T. D. STEWART: The electromotive force produced by the acceleration of metals. In: *Physical Review* 8 (1916), p. 97–116.

[Vol00a] VOLTA, A.: On the electricity excited by the mere contact of conducting substances of different kinds. Letter in French from A. Volta to J. Banks dated March 20, 1800. It was read before the Royal Society in June 26, 1800. In: *Philosophical Transactions* 90 (1800), p. 403–431.

[Vol00b] VOLTA, A.: On the electricity excited by the mere contact of conducting substances of different kinds. In: *Philosophical Magazine* 7 (1800), p. 289–311.

[weba] WEBER, W.: Determinations of electrodynamic measure: concerning a universal law of electrical action, 21st Century Science & Technology, posted March 2007, translated by S. P. JOHNSON, edited by L. HECHT AND A. K. T. ASSIS. Available at: http://www.21stcenturysciencetech.com/.

[webb] WEBER, W.: Determinations of electrodynamic measure: particularly in respect to the connection of the fundamental laws of electricity with the law of gravitation, 21st Century Science & Technology, posted November 2008. Translated by G. GREGORY, edited by L. HECHT AND A. K. T. ASSIS. Available at: http://http://www.21stcenturysciencetech.com/.

[Web39] WEBER, W.: Unipolare Induktion. In: *Resultate aus den Beobachtungen des magnetischen Vereins* III (1839), p. 63–90. Reprinted in Wilhelm Weber's *Werke*, Vol. 3, ed. by H. WEBER. Berlin: Springer 1893, p. 153–175.

[Web41] WEBER, W.: Unipolare Induktion. In: *Annalen der Physik und Chemie* 52 (1841), p. 353–386. Reprinted in Wilhelm Weber's *Werke*, Vol. 3, ed. by H. WEBER. Berlin: Springer 1893, p. 176–179.

[Web46] WEBER, W.: Elektrodynamische Maassbestimmungen – über ein allgemeines Grundgesetz der elektrischen Wirkung. *Abhandlungen bei Begründung der Königl. Sächs. Gesellschaft der Wissenschaften am Tage der zweihundertjährigen Geburtstagfeier Leibnizen's herausgegeben von der Fürstl. Jablonowskischen Gesellschaft Leipzig* 1846, p. 211–378. Reprinted in Wilhelm Weber's *Werke*, Vol. 3, ed. by H. WEBER. Berlin: Springer 1893, p. 25–214.

[Web48a] WEBER, W.: Elektrodynamische Maassbestimmungen. In: *Annalen der Physik und Chemie* 73 (1848), p. 193–240. Reprinted in Wilhelm Weber's *Werke*, Vol. 3, ed. by H. WEBER. Berlin: Springer 1893, p. 215–254.

[Web48b] WEBER, W.: Über die Erregung und Wirkung des Diamagnetismus nach den Gesetzen induciter Ströme. In: *Annalen der Physik und Chemie* 73 (1848), p. 241–256. Reprinted in Wilhelm Weber's *Werke*, Vol. 3, ed. by H. WEBER. Berlin: Springer 1893, p. 255–268.

[Web52a] WEBER, W.: Elektrodynamische Maassbestimmungen insbesondere über Diamagnetismus. In: *Abhandlungen der Königl. Sächs. Gesellschaft der Wissenschaften, mathematischphysische Klasse* 1 (1852), p. 485–577. Reprinted in Wilhelm Weber's *Werke*, Vol. 3, ed. by H. WEBER. Berlin: Springer 1893, p. 473–554.

[Web52b] WEBER, W.: Elektrodynamische Maassbestimmungen insbesondere Widerstandsmessungen. In: *Abhandlungen der Königl. Sächs. Gesellschaft der Wissenschaften, mathematisch-physische Klasse* 1 (1852), p. 199–381. Reprinted in Wilhelm Weber's *Werke*, Vol. 3, ed. by H. WEBER. Berlin: Springer 1893, p. 301–471.

[Web52c] WEBER, W.: Ueber den Zusammenhang der Lehre vom Diamagnetismus mit der Lehre von dem Magnetismus und der Elektricität. In: *Annalen der*

Physik und Chemie 87 (1852), p. 145–189. Reprinted in Wilhelm Weber's *Werke*, Vol. 3, ed. by H. WEBER. Berlin: Springer 1893, p. 555–590.

[Web55] WEBER, W.: Vorwort bei der Übergabe der Abhandlung: *Elektrodynamische Maassbestimmungen, insbesondere Zurückführung der Stromintensitäts-Messungen auf mechanisches Maass*. In: *Berichte über die Verhandlungen der Königl. Sächs. Gesellschaft der Wissenschaften zu Leipzig, mathematisch-physische Klasse* 17 (1855), p. 55–61. Reprinted in Wilhelm Weber's *Werke*, Vol. 3, ed. by H. WEBER. Berlin: Springer 1893, p. 591–596.

[Web62] WEBER, W.: Zur Galvanometrie. In: *Abhandlungen der Königl. Gesellschaft der Wissenschaften zu Göttingen, mathematische Klasse* 10 (1862), p. 3–96. Reprinted in Wilhelm Weber's *Werke*, Vol. 4, ed. by H. WEBER. Berlin: Springer 1894, p. 17–96.

[Web64] WEBER, W.: Elektrodynamische Maassbestimmungen insbesondere über elektrische Schwingungen. In: *Abhandlungen der Königl. Sächs. Gesellschaft der Wissenschaften, mathematisch-physische Klasse* 6 (1864), p. 571–716. Reprinted in Wilhelm Weber's *Werke*, Vol. 4, ed. by H. WEBER. Berlin: Springer 1894, p. 105–241.

[Web71] WEBER, W.: Elektrodynamische Maassbestimmungen insbesondere über das Princip der Erhaltung der Energie. In: *Abhandlungen der Königl. Sächs. Gesellschaft der Wissenschaften, mathematisch-physische Klasse (Leipzig)* 10 (1871), p. 1–61. Reprinted in Wilhelm Weber's *Werke*, Vol. 4, ed. by H. WEBER. Berlin: Springer 1894, p. 247–299.

[Web72] WEBER, W.: Electrodynamic measurements — Sixth memoir, relating specially to the principle of the conservation of energy. In: *Philosophical Magazine* 43 (1872), p. 1–20 and 119–149.

[Web75] WEBER, W.: Ueber die Bewegung der Elektricität in Körpern von molekularer Konstitution. In: *Annalen der Physik und Chemie* 156 (1875), p. 1–61. Reprinted in Wilhelm Weber's *Werke*, Vol. 4, ed. by H. WEBER. Berlin: Springer 1894, p. 312–357.

[Web78] WEBER, W.: Elektrodynamische Maassbestimmungen insbesondere über die Energie der Wechselwirkung. In: *Abhandlungen der Königl. Sächs. Gesellschaft der Wissenschaften, mathematisch-physische Klasse, (Leipzig)* 11 (1878), p. 641–696. Reprinted in Wilhelm Weber's *Werke*, Vol. 4, ed. by H. WEBER. Berlin: Springer 1894, p. 361–412.

[Web92a] WEBER, W.: *Wilhelm Weber's Werke, vol. 2, Magnetismus*. Ed. by E. RIECKE. Berlin: Springer 1892.

[Web92b] WEBER, W.: *Wilhelm Weber's Werke*, W. Voigt, (ed.), vol. 1, *Akustik, Mechanik, Optik und Wärmelehre*. Berlin: Springer 1892.

[Web93] WEBER, W.: *Wilhelm Weber's Werke*, H. Weber (ed.), vol. 3, *Galvanismus und Elektrodynamik*, first part. Berlin: Springer 1893.

[Web94a] WEBER, W.: Elektrodynamische Maassbestimmungen insbesondere über den Zusammenhang des elektrischen Grundgesetzes mit dem Gravitationsgesetze. In: WEBER, H.: *Wilhelm Weber's Werke, Vol. 4*. Berlin: Springer 1894, p. 479–525.

[Web94b] WEBER, W.: *Wilhelm Weber's Werke*, H. Weber (ed.), vol. 4, *Galvanismus und Elektrodynamik*, second part. Berlin: Springer 1894.

[Web66a] WEBER, W.: On the connexion of diamagnetism with magnetism and electricity. In: TYNDALL, J. AND W. FRANCIS (ed.): *Scientific Memoirs, Vol. 7*. New York: Johnson Reprint Corporation 1966, p. 163–199.

[Web66b] WEBER, W.: On the excitation and action of diamagnetism according to the laws of induced currents. In: TAYLOR, R. (ed.): *Scientific Memoirs, Vol. 5*. New York: Johnson Reprint Corporation 1966, p. 477–488.

[Web66c] WEBER, W.: On the measurement of electro-dynamic forces. In: TAYLOR, R. (ed.): *Scientific Memoirs*, Vol. 5. New York: Johnson Reprint Corporation 1966, p. 489–529.

[Whi73] WHITTAKER, E. T.: *A History of the Theories of Aether and Electricity*, vol. 1: *The Classical Theories*. New York: Humanities Press 1973.

[Wie60] WIEDERKEHR, K. H.: *Wilhelm Webers Stellung in der Entwicklung der Elektrizitätslehre*. Dissertation, Hamburg 1960.

[Wie67] WIEDERKEHR, K. H.: *Wilhelm Eduard Weber – Erforscher der Wellenbewegung und der Elektrizität (1804–1891)*. Ed. by H. DEGEN. Stuttgart: Wissenschaftliche Verlagsgesellschaft (Grosse Naturforscher; vol. 32) 1967.

[Wie88] WIEDERKEHR, K. H.: Zur Deutung magnetischer Phänomene im 19. Jahrhundert. In: *Physikalische Blätter* 44 (1988), p. 129–134.

[Wie91] WIEDERKEHR, K. H.: Faradays Feldkonzept und Hans Christian Oersted. In: *Physikalische Blätter* 47 (1991), p. 825–830.

[Wie93] WIEDERKEHR, K. H.: Das Experiment von Wilhelm Weber und Rudolf Kohlrausch 1855 und Maxwells elektromagnetische Lichttheorie. In: SCHRÖDER, W. (ed.): *The Earth and the Universe. A Festschrift in honour of Hans-Jürgen Treder*. Interdivisional Commission on History of the International Association of Geomagnetism and Aeronomy. Bremen-Rönnebeck 1993, p. 452–463.

[Wie94] WIEDERKEHR, K. H.: Wilhelm Weber und Maxwells elektromagnetische Lichttheorie. In: *Gesnerus* 51 (1994), Part 3/4, p. 256–267.

[Wie99] WIEDERKEHR, K. H.: Die Entdeckung des Elektrons. In: *Der Mathematisch-Naturwissenschaftliche Unterricht* (MNU) 52 (1999), p. 132–139.

[Wie04] WIEDERKEHR, K. H.: Ein bisher unveröffentlichter Brief von Rudolf Kohlrausch an Andreas v. Ettingshausen von 1854, das Kohlrausch-Weber-Experiment von 1854/55 und die Lichtgeschwindigkeit in Wilhelm Webers Elektrodynamik. In: *NTM – International Journal of History and Ethics of Natural Sciences, Technology and Medicine* 12 (2004), p. 129–145.

[Wie07] WIEDERKEHR, K. H.: Über Vorstellungen vom Wesen des elektrischen Stromes bis zum Beginn der Elektronentheorie der Metalle. In: WOLFSCHMIDT, G. (ed.): *"Es gibt für Könige keinen besonderen Weg zur Geometrie" – Festschrift für Karin Reich*. Augsburg: Dr. Erwin Rauner Verlag 2007, p. 299–308.

[Wie08] WIEDERKEHR, K. H.: Heinrich Hertz between the old electrodynamics and Maxwell's theory. In: WOLFSCHMIDT, G. (ed.): *Heinrich Hertz (1857–1894) and the Development of Communication*. Norderstedt: Books on Demand (Nuncius Hamburgensis – Beiträge zur Geschichte der Naturwissenschaften; Vol. 10) 2008, p. 151–159.

[WK56] WEBER, W. AND R. KOHLRAUSCH: Über die Elektricitätsmenge, welche bei galvanischen Strömen durch den Querschnitt der Kette fliesst. In: *Annalen der Physik und Chemie*, ed. by J. C. POGGENDOFF 99 (1856), p. 10–25. Reprinted in Wilhelm Weber's *Werke*, Vol. 3, ed. by H. WEBER. Berlin: Springer 1893, p. 597–608.

[WK68] WEBER, W. AND R. KOHLRAUSCH: Über die Einführung absoluter elektrischer Maße. Ed. by BALKE, S.; GERICKE, H.; HARTNER, W.; KERSTEIN, G.; KLEMM, F.; PORTMANN, A.; SCHIMANK, H. AND K. VOGEL. Commented by F. KOHLRAUSCH AND K. H. WIEDERKEHR. Braunschweig: Friedrich-Vieweg & Sohn (Ostwalds Klassiker der exakten Wissenschaften, new series; Vol. 5) 1968.

[WK03] WEBER, W. AND R. KOHLRAUSCH: On the amount of electricity which flows through the cross-section of the circuit in galvanic currents. In: BEVILACQUA, F. AND E. A. GIANNETTO (ed.): *Volta and the History of Electricity*. Translated by S. P. JOHNSON. Milano: Università degli Studi di Pavia and Editore Ulrico Hoepli 2003, p. 287–297. Available at http://www.ifi.unicamp.br/~assis.

[WW93] WEBER, E. H. AND W. WEBER: *Wilhelm Weber's Werke*, E. Riecke (ed.), vol. 5, *Wellenlehre auf Experimente gegründet oder über die Wellen tropfbarer Flüssigkeiten mit Anwendung auf die Schall- und Lichtwellen*. Berlin: Springer 1893. Originally published in 1825.

[WW94] WEBER, W. AND E. WEBER: *Wilhelm Weber's Werke*, ed. by F. MERKEL AND O. FISCHER (ED.), vol. 6, *Mechanik der menschlichen Gehwerkzeuge. Eine anatomisch-physiologische Untersuchung*. Berlin: Springer 1894. Originally published in 1836.

[Zol76] ZÖLLNER, J. C. F.: *Principien einer elektrodynamischen Theorie der Materie*. Leipzig: Engelmann 1876.

[Zol78a] ZÖLLNER, J. C. F.: Ueber die Ableitung der Newton'schen Gravitation aus den statischen Wirkungen der Elektricität. In: ZÖLLNER, J. C. F. (ed.): *Wissenschaftliche Abhandlungen*. Leipzig: L. Staackmann 1878, Vol. 1, p. 417–459.

[Zol78b] ZÖLLNER, J. C. F.: Ueber die elektrischen Wirkungen des Lichtes und der strahlenden Wärme. In: ZÖLLNER, J. C. F. (ed.): *Wissenschaftliche Abhandlungen.* Leipzig: L. Staackmann 1878, Vol. 1, p. 600–610.

[Zol82] ZÖLLNER, J. C. F.: *Erklärung der Universellen Gravitation aus den statischen Wirkungen der Elektricität und die allgemeine Bedeutung des Weber'schen Gesetzes.* Leipzig: L. Staackmann 1882.

[Zol83] ZÖLLNER, J. C. F.: *Über die Natur der Cometen – Beiträge zur Geschichte und Theorie der Erkenntniss.* First edition of 1872. Leipzig: L. Staackmann (3rd edition) 1883.

Figure 2.1:
Wilhelm Eduard Weber (1804–1891) – 1865
Kohlrausch, F. (Oswalds Klassiker Nr. 142) 1904, Frontispiz.

Vorstellungen von der elektrischen Leitung und Entwicklung der Elektronentheorie der Metalle von Riecke, Drude, Lorentz bis Sommerfeld

Karl Heinrich Wiederkehr und Gudrun Wolfschmidt

(Hamburg)

2.1 Einleitung

Für viele Menschen stellt die Fortleitung der Elektrizität in Drähten eine höchst einfache Sache dar. Man weiß, dass André Marie Ampère (1775–1836) von einem Doppelstrom gesprochen hatte und dies dann korrigiert werden musste; denn durch das Metall fließen nur Elektronen, ähnlich wie Wasser durch Röhren. Betrachtet man jedoch die historische Entwicklung, stellt sich heraus, dass die elektrischen Leitungsvorgänge in Metallen lange Zeit kontrovers diskutiert wurden und diese Vorgänge äußerst kompliziert sind. Erst im 3. Jahrzehnt des 20. Jahrhunderts konnten die Vorgänge endgültig geklärt werden. Den Vorstellungen eines Michael Faradays (1791–1867), André Marie Ampères, Wilhelm Webers (1804–1892), James Clerk Maxwells (1831–1879) bis hin zu Arnold Sommerfeld (1868–1951) soll hier nachgespürt werden.

2.2 Michael Faraday (1791–1867) und André Marie Ampère (1775–1836)

Mit der Entdeckung der galvanischen Elektrizität durch Luigi Galvani (1737–1798) und Alessandro Volta (1745–1827) und der Schaffung galvanischer Elemente hatte man Elektrizitätsquellen zur Verfügung, die bei geringer Spannung relativ konstante, große elektrische Ströme lieferten – im Gegensatz zur Reibungselektrizität.

Abbildung 2.2:
Luigi Galvani (1737–1798) und Alessandro Volta (1745–1827)
Lenard 1930, S. 142. Gerrits 1948, S. 225.

1820 entdeckte dann Hans Christian Ørsted (1777–1851) den Elektromagnetismus (vgl. Abb. 1.2, S. 21). Viele Physiker warfen sich sogleich auf dieses neue Gebiet, bei dem man neben galvanischen Elementen nur eine Magnetnadel benötigte.

Der erfolgreichste Forscher war hier zunächst Ampère (vgl. Abb. 1.3, S. 23). In einem atemraubenden Tempo machte er seine Entdeckungen, die er seit September 1820 der Pariser Akademie vortrug. Er führte den elektrischen Doppelstrom ein: das positive elektrische Fluidum floß in der einen Richtung durch den Draht, das negative Fluidum in der entgegengesetzten Richtung. Mit seiner Schwimmerregel, die den Drehsinn der Magnetnadel angab, setzte er die Richtung des (positiven) elektrischen Stromes fest. Begegnen sich die Teilchen der beiden Flüssigkeiten, sollen sich diese zu dem neutralen Fluidum, dem Äther vereinigen und durch die elektromotorische Kraft (Spannung) wieder getrennt werden. In seinem elektrodynamischen Fundamentalgesetz – es ist ein Fernwirkungsgesetz (unendlich große Ausbreitungsgeschwindigkeit der Wirkung ohne vermittelndes Medium) – erfasste er qualitativ die Anziehungs- und Abstoßungskräfte stromdurchflossener Leiter (ohne Zuhilfenahme des Magnetfeldes), und führte mit seiner Molekularstromhypothese allen Magnetismus auf fließende elektrische Ströme zurück.[1]

Faradays Debüt seiner Erforschung der elektrischen Erscheinungen war ebenfalls die Oerstedsche Entdeckung und die Sichtung einer Flut von Arbeiten darüber.[2]

Faradays Vorstellungen wurden mehr von Ørsted beeinflusst, der von einem kreisförmigen elektrischen Konflikt sprach; dieser sollte sich längs des Leiters um den Leiter herum fortsetzen. Faraday (vgl. Abb. 1.4, S. 25) interpretierte die kreisförmigen Linien allerdings als magnetische Kraftlinien. Bei Ørsted spielte romantisch-naturphilosophisches Gedankengut mit hinein, Leitmotive wie Dualität, Kontinuum und Dynamismus, Homogenität und Heterogenität. Mit dem Ampereschen Doppelstrom konnte sich Faraday nicht anfreunden und scheute sich, Elektrizität als Fluidum, als eine imponderable Substanz anzuerkennen und zu behandeln. Bei allen elektrischen Vorgängen war seiner Ansicht nach ein fundamentales Prinzip, die elektrische Influenz, im Spiel. Er nannte es „elektrostatische Induktion". Längs einer Linie, sie entspricht später der elektrischen Feldlinie, soll eine elektrische Kraft auf die Körperteilchen (besser gesagt auf die Volumenelemente des Körpers) einwirken und diese polarisieren. In einem Dielektrikum, einem Elektrolyten und auch in einem metallischen Leiter erfolgt so eine Influenz, eine „Verteilung" der Elektrizität in positive und negative Elektrizität. Während es in einem Dielektrikum bei der „Verteilung" bleibt, erfolgt in einem metallischen Leiter in kürzester Zeit eine Mitteilung, ein Ladungsübergang. Mit der andauernden Wiederholung dieser Vorgänge deutete Faraday den elektrischen Strom. Wie schon gesagt akzeptierte Faraday den Ampereschen Doppelstrom nicht. Nach ihm besteht der elektrische Strom in einer Art Bewegung, einer „Kraft" (es müsste besser Energie heißen), die sich fortpflanzt, wobei der Leitungsdraht nur eine Art „Vehikel" ist.

---

[1] Zu Ampère: Rosenberger 1887–1890. - Pearce Williams, L.: A. M. Ampère. In: Meÿenn, Band 1 (1997), S. 336–356.

[2] Faraday, Michael: Historical Sketch of Electro-Magnetism, zuerst erschienen in *Annals of Philosophy, New Series* Vol. 2 (1821) und Vol. 3 (1822). Faraday: Experimental Researches in Electricity. Deutsche Übersetzung S. Kalischer: M. Faraday Experimental-Untersuchungen über Elektricität, 3 Bde. Berlin 1889–1891, hierzu Bd. 2, S. 265–297. Der Einfachheit halber werden auch die Artikelnummern bei Faraday angegeben.

Wir meinen, daß von Ørsted Anstöße auf Faraday ausgingen, die nicht nur die kreisförmigen magnetischen Feldlinien betreffen. Auch in den linearen dynamischen Vorgängen innerhalb des Leitungsdrahtes, beim Übergang vom homogenen zum heterogenen (Polarisation), sind romantisch-naturphilosophische Elemente erkennbar. Begriffe wie „elektrische Verteilung" und „Mitteilung" findet man auch schon in Ørsteds „Ansicht der chemischen Naturgesetze" 1812. Und aus dem Historical Sketch geht hervor, dass Faraday auch Ørsteds Arbeit von 1812 in der französischen Fassung von Marcel de Serres (1815) gekannt hat.[3]

2.3 James Clerk Maxwell und die „Natur des elektrischen Stroms"

James Clerk Maxwell (1831–1879) hatte sich intensiv mit Faradays Konzept der magnetischen und elektrischen Feldlinien befasst, Faradays Idee der Nahwirkung ganz in sich aufgenommen und diese mit seinem berühmten vier Gleichungen mathematische Gestalt verliehen in seinem epochemachenden Werk, das mit seinem *„Treatise on Electricity and Magnetism"* (1873) abschließt. Hier sollen nur seine Vorstellungen über den elektrischen Strom in einem metallischen Leiter betrachtet werden und seine Auseinandersetzung mit gegnerischen Ansichten, die an Ampère anschließen. Hauptvertreter der letzteren Richtung waren Gustav Theodor Fechner (1801–1887), Verfasser der *„Atomlehre"* (1855) und Wilhelm Weber (1804–1891), dessen Elektrodynamik auf dem Festland drei Jahrzehnte vor dem endgültigen Sieg der Maxwell-Hertzschen Theorie vorherrschte. Auf jene Endwicklungslinie wollen wir später noch näher eingehen. Nur soviel: aus den imponderablen Flüssigkeiten waren elektrische positive und negative Substanzen geworden mit einer atomistischen Struktur. Die elektrischen Teilchen sollen Masse und damit auch Trägheit besitzen, wobei mindestens der einen Sorte der Teilchen eine sehr kleine Masse zukam.

Maxwell hatte sich ganz auf die Seite Faradays geschlagen und die Existenz elektrischer Substanzen und erst recht eine atomistische Struktur abgelehnt. So lehnte er eine Deutung der elektrochemischen Gesetze mit Hilfe der Annahme eines Elektrizitätsatoms ab – obwohl dies nahe lag. So schreibt er zum Beispiel:

> *„Allein so leicht es ist von einer molekularen Ladung zu sprechen, so schwer ist es, die Existenz einer solchen zu begreifen."*

---

3 Wiederkehr, K. H.: Faradays Feldkonzept und Hans Christian Oersted. In: Physikalische Blätter 47 (1991), S. 825–830. - Wiederkehr, K. H.: Oersteds „Ansicht der chemischen Naturgesetze" 1812 und seine Naturphilosophischen Betrachtungen über Elektrizität und Magnetismus. In: Gesnerus 47 (1990), S. 161–183, bes. S. 178–179.

Abbildung 2.3:
James Clerk Maxwell (1831–1879)
Keferstein 1911, S. 158.

Für Maxwell liegt das Wesentliche der Elektrizität nicht in den Substanzen, sondern im magnetischen und elektrischen Feld.

Zur Prüfung und Untermauerung seiner Ansicht, dass der elektrische Strom nicht mit etwas materiellen Fließenden verknüpft ist, schlug Maxwell zwei Experimente vor. Wir wollen nur auf das erste näher eingehen.[4] Eine Spule liegt mit ihren Windungen in einer horizontalen Ebene; sie hängt an einem vertikalen Draht, durch den auch der Strom zugeführt und über ein Quecksilbergefäß abgeführt wird. Der untere Teil des Drahtes taucht in dieses Gefäß ein. Durch Einschalten des angeblich mit Trägheit behafteten Stromes entsteht ein Drehmoment, das auf Grund des Drehimpulserhaltungssatzes eine Drehbewegung

---

4 Maxwell, J. Cl.: Treatise on Electricity and Magnetism (1873). Deutsche Übersetzung, Bd. 1 (1883), S. 396. - Faraday, M.: Experimental Researches, Artikel 1168 (1837) und 1338 (1839). Deutsche Übersetzung: Bd. 2, S. 262–265. Treatise, Artikel 574, Zitat.

und Verdrillung des aufgehängten Drahtes zur Folge hat. Angezeigt wird diese Drehung mittels eines Spiegels am Draht und einem Lichtzeiger. Die Einwirkung der Vertikalkomponente des erdmagnetischen Feldes wurde vorher durch einen Magneten kompensiert. Beim Ausschalten des Stromes müsste dann eine entgegengesetzte Drehung wahrgenommen werden. Ein derartiger Effekt zeigte sich nicht und Maxwell fühlte sich bestätigt. Dennoch will er nicht generell einen positiven Ausgang eines verfeinerten Experimentes zu Gunsten der gegnerischen Seite ausschließen. Er weiß nichts über die Masse und die Geschwindigkeit der fließenden elektrischen Materie, und vielleicht ist sein Apparat auch nicht empfindlich genug. Hier zeigt sich die ganze Größe des unvoreingenommenen Forschers, der auch um die Unvollkommenheit seinen eigenen Ansichten und Theorien weiß. Am Ende dieses Artikels (574) schreibt Maxwell:

> „Erst wenn wir einen sicheren Beweis für die materielle Beschaffenheit der Electricität besitzen, wird sich die dynamische Theorie der Electricität in Vollständigkeit bearbeiten lassen. Man wird dann die Wirkungen der Electricität nicht mehr wie hier in diesem Werk noch geschehen muß einem unbekannten etwas, das nur den allgemeinen Gesetzen der Dynamik folgt, zuzuschreiben brauchen. Man wird sie aus bekannten Bewegungen bekannter Teile von Materie ableiten, und nicht wie das ebenfalls hier noch geschehen muß bei dem _Totaleffect_ und den Schlussresultaten stehen bleiben, sondern den ganzen _inneren_ Mechanismus und alle Details der Bewegung der Untersuchung unterwerfen können."

Mit dem Stewart-Tolman-Effekt (1914, 1916), in dem die Trägheit der Elektronen und ihre Existenz in metallischen Leitern experimentell nachgewiesen wurde, und mit der Elektronentheorie der Metalle, die im zweiten so erfolgreichen Anlauf die Mittel der Quantentheorie benutzt, wurden Maxwells Worte Wirklichkeit.

2.4 Gustav Theodor Fechner (1801–1887) und Wilhelm Weber (1804–1891)

Wenden wir uns den Physikern zu, die Anhänger einer anderen Auffassung von dem Wesen der Elektrizität waren als Faraday und Maxwell, und nach einem anderen Paradigma in der Erforschung der Elektrizität vorgingen, – zu G. T. Fechner und W. Weber. 1846 erschien Webers Arbeit „Elektrodynamische Maassbestimmungen insbesondere über ein allgemeines Grundgesetz der elektrischen Wirkung". Dieses Gesetz war ihm Grundlage und Richtschnur für alle

Abbildung 2.4:
Gustav Theodor Fechner (1801–1887)
(Wikipedia)

seine weiteren Forschungen.[5] Das Webersche Gesetz ist ein Wechselwirkungs-

---

5 Das Grundgesetz der elektrischen Wirkung lautet (unter Einschluß einer 1852 durchgeführten Präzisierung):

$$F = \frac{e_1 e_2}{r^2}\left(1 - \frac{1}{c_w^2}\frac{dr^2}{dt^2} + \frac{2r}{c_w^2}\frac{d^2r}{dt^2}\right)$$

F Kraft, e elektrische Ladung, r Abstand der Ladungen, $\frac{dr}{dt}$ relative Geschwindigkeit der bewegten Ladungen e_1 und e_2, $\frac{d^2r}{dt^2}$ relative Beschleunigung, c_w Webersche Konstante ($c_w = \sqrt{2}c_L$; c_L Lichtgeschwindigkeit im Vakuum. Zum Weberschen Gesetz siehe Wiederkehr, K. H.: Wilhelm Eduard Weber, Erforschung der Wellenbewegung und der Elektrizität. Stuttgart 1967, hierzu S. 90 ff. - Wiederkehr, K. H.: Carl Friedrich Gauß (1777–1855) und Wilhelm Weber (1804–1891). In: Meÿenn 1997, S. 356–370.

gesetz zwischen zwei elektrischen Ladungen, das nicht nur die elektrostatische Kraft erfasst, sondern auch die elektrodynamische Kraft zwischen zwei bewegten elektrischen Ladungen. Dabei wird die relative Geschwindigkeit (Abstandsänderung) zwischen den beiden Ladungen und die relative Beschleunigung zu Hilfe genommen. Sein Gesetz hatte Weber aus dem elektrodynamischen Fundamentalgesetz von Ampère (1822) hergeleitet, das die Kraft zwischen zwei Stromelementen angibt (Ids Stromelement, I Stromstärke, ds Länge des kleinen Stückchens).

Mit dem Weberschen Gesetz konnte aber nicht nur die Kraft zwischen stromdurchflossenen Leitern erfasst werden, sondern auch die elektromagnetische Induktion. Alle damals bekannte elektrische Phänomene konnte Weber so deuten. Sein Gesetz beherrschte drei Jahrzehnte lang fast die ganze Elektrodynamik auf dem Festlande. Es war auch ein Fernwirkungsgesetz, d. h., die Kraftwirkungen pflanzten sich ohne Zwischenmedium mit einer unendlich großen Geschwindigkeit fort. Im Gegensatz dazu stand die Faraday-Maxwellsche Theorie, die eine Nahwirkung annimmt (Fortpflanzung mit einer endlichen Geschwindigkeit); vermittelndes Medium war der Äther. Mit der Entdeckung der elektromagnetischen Wellen 1888 durch Heinrich Hertz (1857–1894) war der endgültige Sieg über die ältere Elektrodynamik mit ihren Fernwirkungen errungen, ihre Auffassung über eine atomistische Struktur der Elektrizität wurde später wieder aufgenommen.

Wilhelm Webers Abhandlung von 1846 wurde vorbereitet und angekündigt durch eine Arbeit von Gustav Theodor Fechner (1801–1887) in Poggendorffs Annalen 1845 mit dem Titel „Über die Verknüpfung der Faradayschen Inductionserscheinungen mit den Ampereschen electrodynamischen Erscheinungen".[6]

Fechner leistete mit seiner Abhandlung *„Über die Verknüpfung ..."* Vorarbeit für Webers Abhandlung von 1846 und kündigte sie an. Er zerlegte das Stromelement in zwei, dem Betrage nach gleich groß geladene, positive und negative Teilchen, die sich mit gleichgroßer, aber entgegengesetzter Geschwindigkeit längs einer geraden Linie im Leiterelement ds bewegen. Fechner konnte sowohl die Amperesche Anziehung und Abstoßung elektrischer Ströme erklären, als auch die von Faraday 1831 entdeckte Induktion. Fechners Annahme symmetrischer Verhältnisse wurde von W. Weber übernommen. Der symmetrische Doppelstrom in einem Leiter war ein Schwachpunkt in der Weberschen Elektrodynamik und sollte sich später als irrig erweisen. Fechner war mit den drei Weber-Brüdern, Wilhelm Eduard, Eduard Friedrich und Ernst Heinrich eng befreundet, und in den Leipziger Jahren fand ein intensiver Gedankenaustausch zwischen Fechner und Wilhelm Weber statt, und damit auch ein Einfluß

---

6 Annalen der Physik und Chemie, hrsg. von J. C. Poggendorff, (2) 64 (1845) S. 337.

von Fechners Seite aus. James Clerk Maxwell bemerkte in dem Artikel 475 in seinem „*Treatise*", dass nach der Fechnerschen Hypothese vom symmetrischen elektrischen Doppelstrom kein Effekt in seinem Experiment zu erwarten sei. Die Bewegung einer elektrischen Substanz darf nur in einer Richtung erfolgen.

Im Grundgesetz der elektrischen Wirkung, kurz Webersches Gesetz genannt, wird die Bewegung der elektrischen Teilchen als geradlinig und parallel zum Leitungsdrahtstückchen gesetzt, – und dies ist nach Weber eine Idealisierung. In Wirklichkeit sind die Bewegungen viel komplizierter. Im Leiter finden die Teilchen einen Widerstand, und daraus folgt das Ohmsche Gesetz. Der Atomistiker konnte sich damit nicht zufrieden geben. Er musste das „Wesen" des Widerstandes vom molekularen Bilde her deuten, auch wenn dies nach dem damaligen Forschungsstande nur auf spekulativem Wege möglich war. In den „Elektrodynamischen Maassbestimmungen, insbesondere Widerstandsmessungen" (1852) – es ist die zweite in dieser, sieben Abhandlungen umfassenden Reihe – will er den Ursachen des Widerstandes nachgehen. Die ponderablen Atome im Leiter können nicht der alleinige Grund sein. Es müssen noch andere Kräfte da sein, die einen gleichförmigen, beharrlichen Strom hervorbringen und der fortwährenden Spannung (elektromotorische Kraft) das Gleichgewicht halten. Weil Weber die Ampereschen Molekularströme beibehalten will – er braucht diese bei seiner Theorie des Magnetismus und Diamagnetismus, entsteht nach ihm der Widerstand in einem metallischen Leiter durch eine sich andauernd wiederholende Verbindung und Scheidung positiver und negativer elektrischer Teilchen. Zur Veranschaulichung denkt sich Weber die punktförmigen positiven Teilchen in regelmäßigen Abständen auf einer geraden Linie angeordnet. Sie sind an einen festen Ort gebunden und mit den ponderablen Atomen verschmolzen. Es sei bemerkt, daß hier W. Weber vom elektrischen Doppelstrom zum Einfachstrom übergeht. Um diese positiven Teilchen kreisen die negativen elektrischen Teilchen. Wird eine Spannung angelegt, werden die Kreise zu Ellipsen. Die negativen Teilchen verlassen auf Spiralbahnen ihr positives Zentrum, werden beschleunigt und geraten in die Wirkungssphäre des benachbarten positiven Zentrums, und das Spiel beginnt immer wieder von neuem. Nach Weber liegt der Widerstand des Leiters in den Zentralkräften der positiven Ladungen. Die Anzahl der Scheidungen, bzw. Übergänge in einer bestimmten Zeit ist der Stromstärke proportional – und in Verbindung mit der Spannung ergibt sich der Widerstand.

Wilhelm Weber vervollkommnte und verfeinerte Ampères Molekularstromhypothese, indem er für den Magnetismus des Eisens und des Stahls drehbare Elementarmagnete (Kreisströme) annahm, die sich in einem Magnetfeld ausrichten. Den Diamagnetismus deutete er mit der Induktion von Molekularströmen beim Hineinbringen des diamagnetischen Körpers in ein Magnetfeld, und

Abbildung 2.5:
Tangentenbussole
Das Magnetfeld einer um die Kompaßnadel herumgeführten, vom Strom durchflossenen Kreiswindung lenkt die Nadel ab. Mit diesem Instrument führte Wilhelm Weber die erste absolute elektromagnetische Strommessung durch.
Foto: Gudrun Wolfschmidt

die dann beim Herausziehen durch abermalige Induktion wieder verschwinden. Es waren erste Grundlagen für die heutigen Theorien.[7]

Wie erklärt sich Wilhelm Weber das Vorhandensein von Nichtleitern, Isolatoren? Auch in diesen sind Amperesche Molekularströme vorhanden; aber bei ihnen finden keine Wurfbewegungen elektrischer Teilchen von einer Wirkungssphäre in die andere statt. In allen Körpern sind nach Weber Amperesche Molekularströme vorhanden.

Der Wärmeinhalt von Leitern und Isolatoren ist nach W. Weber identisch mit der Bewegungsenergie der elektrischen Teilchen. Die Wärmefortpflanzung geschieht bei metallischen Leitern durch die Emission der schnelleren Teilchen von den wärmeren Wirkungssphären in die kälteren. Was geschieht bei den Isolatoren? W. Weber sagt, die Wärmeverbreitung findet hier mittels Wärmestrahlung statt, die der Lichtstrahlung ähnlich ist.[8] Angeregt zu seiner These der Wärmestrahlung in Isolatoren wurde Wilhelm Weber durch die Dissertation von Carl Neumann (1832–1925), mit dem Titel „*Explicare tentatur quomodo fiat ut lucis polarisationis per vires electricas vel magneticas declinetur*" (Halis saxonum, 1858).[9]

Carl Neumann (vgl. Abb. ??, S. ??) ist der Sohn von Franz Neumann (1798–1895), der mit seinen Arbeiten zur Induktion zusammen mit W. Weber die ältere Elektrodynamik maßgeblich gestaltete.[10] Hendrik Antoon Lorentz, der Schöpfer der Elektronentheorie Ende des 19. Jahrhunderts sprach sein Erstaunen darüber aus, wie nahe seine eigene Deutung des Faraday-Effektes (Drehung der Polarisationsebene des Lichts im Magnetfeld) zu derjenigen Carl Neumanns, eines Verfechters der Weberschen Elektrodynamik, komme.[11] Wilhelm Weber geht aber über C. Neumann hinaus und entwickelt dabei ein Atommodell, bei dem die eine Elektrizitätsart, besser ein elektrisches Teilchen, wieder mit der ponderablen Masse eines Atoms oder Moleküls verschmolzen ist, und der Amperesche Molekularstrom aus der anderen Elektrizitätsart, hier aus der negativen, aus rotierenden negativen Teilchen, besteht. Durch diese Rotationsbewegung

---

[7] Weber, Wilhelm: Elektrodynamische Maassbestimmungen, insbesondere Widerstandsmessungen. (1852). Weber Werke, Bd. 3, S. 391–471, bes. 401–405.

[8] Weber, Wilhelm: Zur Galvanometrie (1862). In: Weber Werke, Bd. 4, S. 17–96, bes. S. 91–96. - Weber Werke Bd. 4, S. 343. Über die Bewegungen der Elektricität in Körpern von molekularen Konstitution (1875), S. 312–357; Annalen der Physik und Chemie, hrsg. von J. C. Poggendorff, (2) 156 (1875), S. 1–61.

[9] Deutsch: Die magnetische Drehung der Polarisationsebene des Lichtes (Halle 1863).

[10] Siehe dazu auch Wiederkehr: W. Weber und die Entwicklung der Elektrodynamik. In: II. Webersymposium 1993, S. 39–54, bes. S. 51.

[11] Siehe dazu Lorentz, H. A.: Theorie der magneto-optischen Phänomene (1909). In: Encyklopädie der Mathematischen Wissenschaften, 5. Bd. (Physik), 3. Teil, Leipzig 1909–1926, S. 256.

werden in der angrenzenden Ätherschicht Wellen der Wärmestrahlung (ähnlich wie die Lichtwellen) erregt. Weber schreibt:[12]

> „*Findet dann aber wirklich eine Störung des Gleichgewichts in der unmittelbar angrenzenden Ätherschicht, folglich eine Erregung von Ätherwellen, statt, so leuchtet ein, dass dieselbe mit jedem Umlauf der Elektricität um das Molekül sich wiederhole, also die Wellendauer mit der Umlaufszeit der elektrischen Teilchen im Molekularstrom übereinstimmen muß. Bei leuchtenden Molekülen ist aber die Wellendauer der von ihnen ausgesandten Wellenzügen aus optischen Versuchen genau bekannt; es würde also, wenn die angenommene Relation zwischen elektrischen Molekularströmen und dem Lichtäther, nach Neumanns Idee sich bestätigte, hier nach möglich werden, aus optischen Versuchen über das Verhalten der die Molekularströme bildenden Elektricität nähere Auskunft erhalten.*"

In seinen „*Elektrodynamischen Maassbestimmungen, insbesondere über das Princip von der Erhaltung der Energie*" (1871) gibt Weber gegen Schluß der Abhandlung auch eine erste Deutung des Thermomagnetismus mit Hilfe der Bewegungsenergie der Molekularströme.[13] Da die Bewegungsenergie von der Masse der Teilchen und deren Geschwindigkeit abhängt, kann bei gleicher Temperatur (gleichgroße Bewegungsenergie) bei dem einen Leiter eine größere Masse mit kleinerer Geschwindigkeit rotieren als in dem anderen Leiter mit kleinerer Masse und größerer Geschwindigkeit. Wilhelm Weber will damit den von Jean Charles Athanase Peltier (1785–1845) und nach ihm benannten Effekt (1834) erklären, nämlich die Erwärmung bzw. Abkühlung an den Kontaktstellen zweier verschiedener Metalle, und ebenso den vom Thomas Johann Seebeck (1770–1831) an Wismut und Kupfer 1821 entdeckten Thermoeffekt (Thermoelement).

Friedrich Kohlrausch (1840–1910), Sohn von Rudolf Kohlrausch (1809–1858), befasste sich etwas genauer mit den Problemen der Thermoelektrizität und Berührspannungen.[14] Wilhelm Weber und Rudolf Kohlrausch entdeckten 1856 bei der Bestimmung des Verhältnisses von absolut elektrostatisch und absolut elektromagnetisch gemessener Elektrizitätsmenge die Lichtgeschwindigkeit. Friedrich Kohlrausch studierte bei Wilhelm Weber in Göttingen und bei Wilhelm Beetz (1822–1886) in München. Weber war ihm ein väterlicher Freund, bei ihm erlernte er die Feinheiten magnetischer und elektrischer Messungen.

---

[12] Weber, Wilhelm: Zur Galvanometrie (1862). In: Weber Werke, Bd. 4, S. 95–96.
[13] Weber Werke, Bd. 4, S. 247–299, bes. S. 294.
[14] Kohlrausch 1875. Wiederkehr (1994). Wiederkehr (2004).

Abbildung 2.6:
Friedrich Kohlrausch (1840–1910)
http://en.wikipedia.org/wiki/File:Friedrich_Kohlrausch.jpg

In München wurde Friedrich Kohlrausch zu den Leitfähigkeitsmessungen und zur Ionentheorie hingeführt. Von 1866 an organisierte Kohlrausch das physikalische Praktikum in Göttingen und schrieb 1870 den bekannten „*Leitfaden der praktischen Physik*" (später „Praktische Physik", bekannt als „der Kohlrausch"). Zuletzt war er Präsident der Physikalisch-Technischen Reichsanstalt in Berlin-Charlottenburg.

Die Arbeit „Über Thermoelektricität, Wärme und Elektricitätsleitung" (1875) enthält seine „Mitführungstheorie". Nach dieser ist mit dem elektrischen Strom eine Mitführung von Wärme verknüpft und andrerseits mit dem Wärmestrom ein elektrischer Strom verknüpft. In dem von Gustav Wiedemann (1826–1899) und dem Gymnasiallehrer Rudolph Franz 1853 entdeckten Gesetz (besser Regel) über die Proportionalität von Wärmeleitfähigkeit und elektrischer Leitfähigkeit sah Kohlrausch einen Hinweis und eine Stütze für seine Mitführungstheorie. Das Wiedemann-Franzsche Gesetz sagt aus, dass in guter Nährung für alle Metalle der Quotient $\frac{\kappa}{\sigma}$ (κ thermische Leitfähigkeit, σ elektrische Leitfähigkeit) eine Konstante ist. Der dänische Physiker Ludwig Lorenz (1829–1891) fügte diesem Gesetz 1872 eine Temperaturabhängigkeit hinzu ($\frac{\kappa}{\sigma} \times LT$; wobei L Lorenzkonstante, T absolute Temperatur ist).

Von einer Wärmefortführung durch den elektrischen Strom hatte auch schon William Thomson, Lord Kelvin (1824–1907) 1856 gesprochen. Er wollte damit den von ihm entdeckten Effekt erklären, dass das Joulesche Erwärmungsgesetz eines stromdurchflossenen Leiters bei unterschiedlichen Temperaturen längs des Leiters nicht streng gültig ist.[15]

2.5 Die klassische Elektronentheorie der Metalle

Was war im letzten Jahrzehnt des 19. Jahrhunderts die gängige, von den meisten Physikern akzeptierte Vorstellung über das Wesen des elektrischen Stromes? Schauen wir in das weitverbreitete Werk „Die Lehre von der Electricität", verfasst von einem der damals einflussreichsten Physiker, Gustav Wiedemann. Im letzten Band seines vierbändigen Werkes, erschienen 1898 werden unter dem Kapitel Bewegungen der Elektrizität zunächst die Nichtleiter (Isolatoren) behandelt. In jedem einzelnen Molekül wird die neutrale Elektrizität durch die angelegte Spannung in positive und negative Elektrizität geschieden. Ein Übergang der Elektrizität von einem Molekül zu einem anderen Molekül findet nicht statt. Zur Leitung in Metallen liest man dann:

> „Endlich findet noch zwischen den durch eine äußere elektrische Scheidungskraft polarisierte Molekülen eine allmähliche Ausgleichung der einander benachbarten, entgegengesetzten Elektrizitäten statt, die innere Ladung verschwindet, und der Körper behält nur an seinem Ende die Ladung. Dies ist der Vorgang der Leitung, welcher bei den vollkommenden Leitern sehr schnell vor sich geht. Dieser Elektrizitätsbewegung steht ein gewisser Widerstand entgegen . . . ."

Das Phänomen elektrischer Strom wird also von Wiedemann noch ganz im Sinne von Faraday und Maxwell gedeutet. Der Nachweis der elektromagnetischen Wellen durch Heinrich Hertz (1857–1894) 1888 mag dazu wohl hier beigetragen haben. Oder besser: Wiedemann stand hier wohl noch ganz unter dem Ein-

---

15 Thomson, William In: Trans. Roy. Soc. London 1856, S. 655.

druck der Hertzschen Entdeckung und dem Triumph der Faraday-Maxwellschen Theorie.[16]

Die klassische Elektronentheorie der Metalle wurde im Wesentlichen von drei Physikern geschaffen:

- Eduard Riecke (1845–1915) (vgl. Abb. 2.7, S. 119),
- Paul Drude (1863–1906) (vgl. Abb. 2.9, S. 124) und
- Hendrik Antoon Lorentz (1853–1928) (vgl. Abb. 2.10, S. ??).

Nach Riecke soll Wilhelm Giese Vorarbeit geleistet haben.[17] Giese will alle elektrischen Leitungsvorgänge von der Elektrolyse über die Leitung durch Flammen bis hin zur metallischen Leitung unter einen Gesichtspunkt bringen. Er weiß, dass er mit seiner Hypothese von substantiellen Ladungsträgern auf den Widerspruch vieler Zeitgenossen stößt. Diese glauben, dass auch noch das letzte verbliebene Imponderable, die elektrischen Flüssigkeiten, nach den Lichtteilchen, der Wärmesubstanz und den magnetischen Fluida verschwindet und in irgendeiner Bewegungsform des Äthers aufgehen wird. Giese beruft sich dabei auf das von Hermann von Helmholtz (1821–1894) 1882 gefundene elektrische Elementarquantum. Die Annahme von Ionen, elektrisch geladenen Bruchstückchen der Moleküle, erwiesen sich bei den Elektrochemikern als besonders fruchtbar. Die Leitung in Flammengasen soll nach Giese ebenfalls durch Ionen geschehen und auch in Metallen, z.B. in Kupfer, sollen Cu^+- und Cu^--Ionen existieren. Ein neutrales Kupfermolekül soll aus so einem Ionenpaar bestehen. Im Gegensatz zur Elektrolyse findet nach Giese bei der Elektrizitätsleitung durch die Ionen in Metallen keine Bewegung der Ionen statt. Sie sind fest an einen Ort gebunden. Die Elektrizitätsleitung erfolgt durch Spaltung und Neubildung der Moleküle unter dem Einfluß der äußeren Spannung:

$$Cu^+ + (Cu^- Cu^+) \rightarrow (Cu^+ Cu^-) + Cu^+$$

Giese hat also noch keine freien beweglichen elektrischen Teilchen, wie z.B. die Elektronen; die elektrische Elementarladung, ist immer an einen ponderablen Teil gebunden.

2.6 Eduard Riecke (1845–1915)

Die klassische Elektronentheorie beginnt mit der Abhandlung von Eduard Riecke (1845–1915); sie hat den Titel „Zur Theorie des Galvanismus und der Wärme"

---

16 Wiedemann 1895–1898, Bd. 4, S. 808–809.
17 Giese, W.: Grundzüge einer einheitlichen Theorie der Electricitätsleitung. In: Annalen der Physik und Chemie, hrsg. von G. Wiedemann, N.F., Bd. 37 (1889), S. 576–609.

(1898).[18] Riecke war 1881 in Göttingen Nachfolger von Wilhelm Weber auf dem Lehrstuhl für Physik. Weber war sein eigentlicher Lehrer gewesen.[19] Riecke erweiterte hier die Ideen W. Webers in einer umfangreichen mathematischen Darstellung. In späteren Abhandlungen[20] von 1907, 1909 und 1915 nahm er auch Gedanken und Ergebnisse von Paul Drude (1863–1906), Hendrik Antoon Lorentz (1853–1928) und anderen auf und konnte so zu seiner Zeit eine in sich geschlossene Darstellung dieses Wissensgebietes geben.

Werfen wir zuerst einen Blick auf den Forschungs- und Wissensstand im letzten Jahrzehnt des 19. Jahrhunderts. Pieter Zeeman (1865–1943) hatte 1896 die magnetische Aufspaltung der Natriumlinie entdeckt. Lorentz konnte sie mit Hilfe von Elektronen – anfangs sprach er noch von Ionen –, die sich auf Kreisbahnen bewegten und unter dem Einfluß des Magnetfeldes ihre Umlauffrequenz änderten, deuten. Joseph John Thomson (1856–1940) identifizierte 1897 die Kathodenstrahlen als schnellfliegende Korpuskeln, die eine bestimmte elektrische Ladung e und die Masse m_e hatten. Sie wurden nach George Johnstone Stoney (1826–1911) Elektronen genannt.[21]

Die Idee von der Existenz solch elektrischer Atome hatte, wie vorher schon erwähnt, viele Jahre davor Wilhelm Weber ausgesprochen. 1881 sah Riecke in den Kathodenstrahlen solche Weberschen Atome. Er konnte die Kreis- und Spiralbahnen der Kathodenstrahlen bei Einwirkung eines Magnetfeldes deuten und gab sogar die Formel für den Radius der Kreisbahn an. Sie wurde bei den $\frac{e}{m}$-Messungen herangezogen.[22] Wenden wir uns nun der Abhandlung Rieckes von 1898 zu.[23] Nach ihm ist ein ponderables Atom von rotierenden positiven und negativen Teilchen umgeben, die sich in dem umgebenden Raum ergießen. Sie bewirken wie bei Weber auch die Wärmefortpflanzung im Leiter. An wärmeren Stellen haben die Teilchen eine größere Geschwindigkeit und fliegen zu den kühleren Stellen hin. Positive und negative Teilchen haben dem Betragen nach gleichgroße elektrische Ladungen. Ihre Masse kann aber zunächst verschieden sein. Positive elektrische Teilchen können an die ponderablen Moleküle gebunden sein, und damit wird die Anzahl der beweglichen positiven Teilchen kleiner als die Zahl der negativen Teilchen. Die Geschwindigkeit der Teilchen ist pro-

---

18 Annalen der Physik und Chemie, N. F., hrsg. von Gustav und Eilhard Wiedemann. (3) 66 (1898), S. 53–389 und 545–581. Auf S. 1191 bis 1200 ist ein Nachtrag, wo Riecke einen Fehler berichtigt, auf den er durch einen Brief von E. van Everdingen aufmerksam gemacht wurde. Van Everdingen befasste sich insbesondere mit galvanomagnetischen und thermomagnetischen Phänomenen an Wismut.
19 Voigt (1915). - Schmid 1919, S. 83–93.
20 Riecke 1907, S. 24–47. Riecke (1909). Riecke 1915.
21 Siehe dazu Wiederkehr (1999), S. 131–139.
22 Riecke (1881), S. 191–194.
23 Riecke (1898).

Abbildung 2.7:
Eduard Riecke (1845–1915)
VOIGT, WOLDEMAR: Eduard Riecke als Physiker.
In: *Physikalische Zeitschrift* XVI (1915), S. 219–221, Frontispiz.

portional $\sqrt{T}$, T ist die absolute Temperatur. Damit wird die kinetische Energie proportional zu T. Bei seiner Rechnung benutzt Riecke eine mittlere Weglänge; es ist die Flugstrecke zwischen zwei benachbarten Molekülen. Obwohl Riecke die elektrischen Teilchen mit Gasmolekülen vergleicht, berücksichtigt er keine Zusammenstöße zwischen den Teilchen untereinander. Offenbar ist bei ihm die Dichte dieser Teilchen sehr viel geringer als die Dichte der Metallmoleküle oder Metallatome. Mit dem Wärmestrom ist nach Riecke ein, wenn auch sehr kleiner, elektrischer Strom verknüpft und er gibt eine Mitführungszahl für die Elektrizität an. Die rechnerische Durchführung der elektrischen Leitung ist komplizierter als bei der Wärmeleitung. Er betreibt eine Art Geometrie der Teilchenbahnen. Sie sind nicht geradlinig, sondern gekrümmt durch den Einfluss der angelegten Spannung. Der galvanische Strom führt auch einen kleinen Wärmestrom mit sich, und Riecke gibt auch hier einen Mitführungskoeffizienten für die Wärme an. Riecke folgt hier also der wechselseitigen Mitführungstheorie von Friedrich Kohlrausch (1875), auf den er sich auch beruft. Er ergänzt aber diese Theorie durch eine quantitative Betrachtung und einer detaillierten Rechnung mit molekularen Größen. Für die spezifische elektrische Leitfähigkeit eines Metalls findet Riecke die Formel:

$$\gamma = \left(\varepsilon_p P_0 u + \varepsilon_n N_0 v \right) \cdot (1 + \alpha t)$$

Dabei ist ε_p die Ladung der positiven beweglichen elektrischen Teilchen, ε_n die Ladung der negativen beweglichen elektrischen Teilchen, P_0 und N_0 ist die Anzahl der Teilchen pro Volumeneinheit, u und v die Beweglichkeit der Teilchen. Die runde Klammer drückt aus, dass die Anzahl der beweglichen Teilchen nach Riecke mit steigender Temperatur wächst. α ist eine materialbedingte Konstante.

Im Elektrizitätsband von Robert Wichard Pohl findet man bei der Herleitung des Ohmschen Gesetzes für ionisierte Luft einen ähnlichen Ausdruck für die spezifische elektrische Leitfähigkeit. Nur fehlt natürlich hier der Koeffizient in der runden Klammer. Bei Riecke ist $\mid \varepsilon_p \mid = \mid \varepsilon_n \mid$. Wenn Riecke auf S. 373, 374 und 375 auf drei verschiedene Art die Menge der positiven Teilchen Q_P berechnet, welche in einer Sekunde durch die Flächeneinheit in Richtung des galvanischen Stromes hindurchgeht, und die Ergebnisse zum Teil unterschiedlich sind, wird offenbar, mit welchen Schwierigkeiten Riecke hier zu kämpfen hat. Unter zu Hilfenahme seiner Ergebnisse für die thermische Leitfähigkeit k und der elektrischen Leitfähigkeit λ sucht er nach einer Beziehung zwischen beiden (S. 379) und bildet den Quotienten $\frac{\kappa}{\lambda}$. Nach dem Wiedemann-Franz-Lorenzschen Gesetz stellt dieser Quotient für alle Metalle in guter Näherung eine Konstante dar. Rieckes Ausdruck dafür ist viel zu kompliziert und er kann nicht zu diesem wichtigen Gesetz kommen. Erst in seiner Abhandlung von 1907 sollte ihm dies unter zu Hilfenahme der Drudeschen gaskinetischen Behandlung der Elektronen gelingen. Seine Formeln erwiesen sich als recht anpassungsfähig.

Peltier fand 1834, dass ein galvanischer Strom, welcher durch ein Thermoelement geleitet wird, an der Lötstelle eine Temperaturveränderung (Abkühlung) hervorbringt, welche derjenigen entgegengesetzt ist, die ein Thermostrom von gleicher Richtung erzeugen würde. In enger Verbindung mit den Thermokräften und dem Peltiereffekt steht der von William Thomson (Lord Kelvin) 1854 beobachtete Effekt. Leitet man einen elektrischen Strom durch einen ungleich erwärmten Metallstab, den man an einem Ende erhitzt, und am anderen Ende abkühlt, wird die Temperaturverteilung längst des Leiters verschieden, je nach der Stromrichtung. Bei der Deutung griff W. Thomson auf die Thermodynamik zurück.

Wie Friedrich Kohlrausch will auch Riecke nun am Schluß seiner Abhandlung 1898 den Peltiereffekt erklären, die thermoelektrischen Gesetzmäßigkeiten und auch den Thomsoneffekt.[24]

Riecke betonte noch einmal, dass die Ladungen der positiven Teilchen und deren Massen in allen Metallen gleich groß sind (ein Teil der positiven Teilchen ist an die ponderablen Atome gebunden). Die negativen Teilchen sind in allen Metallen von gleicher Art und identisch mit den Teilchen der Kathodenstrahlen. Die Ladungen der positiven Teilchen und negativen Teilchen sind ihrem Betrage nach, wie schon gesagt, gleich groß (Elementarquanten). Die Dichte beider Teilchenarten hängt von dem jeweiligen Metall ab, ebenso die Geschwindigkeiten und die freien Weglängen. An den Berührflächen zweier Metalle spricht Riecke auch von einer Diffusion der Teilchen und erfasst die Vorgänge auch rechnerisch. Die Kontaktspannung ist proportional der absoluten Temperatur an der Berührfläche. Bei der Berechnung der thermoelektrischen Spannung muß nach Riecke auch die Mitführung miteinbezogen werden und die daraus resultierende Spannung ist zu der Kontaktspannung zu addieren.

Am Schluß der Abhandlung von 1898 wendet sich Riecke dem galvanomagnetischen und thermomagnetischen Wirkungen zu.[25] Im Ganzen kannte man damals acht Effekte dieser Art (vier transversale und vier longitudinale Effekte) zu. Wir beschränken uns auf die transversalen Effekte. Transversal bedeutet, dass das von außen einwirkende Magnetfeld auf der Richtung des Stromes der durch eine kleine Platte aus einem bestimmten Material fließt, senkrecht steht. Der Strom kann ein Wärmestrom oder ein elektrischer Strom sein. Meist wurden Plättchen aus Wismut benutzt, weil hier die Effekte recht groß sind. Am

---

24 Peltier 1834. Thomson, William: 1854.
25 Riecke 1898, S. 559 und 563. Seeliger 1921, S. 812.

bekanntesten und vielleicht auch am Wichtigsten ist der 1879 von dem amerikanischen Physiker Edwin Hall (1855–1938) an einem Goldplättchen entdeckten Effekt (Halleffekt).[26]

Abbildung 2.8:
Edwin Herbert Hall (1855–1938)
(Wikipedia)

Als Hall an zwei an den Rändern der Folie einander gegenüberliegender Punkte ein empfindliches Galvanometer anschloss, stellte er einen Strom fest. Dieser wird durch eine kleine Spannung (Hall-Spannung) hervorgerufen. Die Richtung

---

26 Biografie zu Edwin H. Hall, siehe Hoffmann et al.: Lexikon der bedeutenden Naturwissenschaftler, Band 2 (F–Mei) 2004, S. 151, Verfasser Reinald Schröder (RS). Bridgeman, P. W.: Biographical Mem. Nation. Acad. Sci. USA 21 (1939), S. 73–94).

des elektrischen Stroms (längs des Plättchens) und der Vektor des magnetischen Feldes stehen senkrecht aufeinander, und die Querspannung, erzeugt durch ein elektrisches Feld mit dem Vektor $\vec{E}$ steht ebenfalls senkrecht auf der durch die vorherigen beiden Vektoren aufgespannten Ebene (rechtwinkliges Dreibein). Gedeutet wird dieser Effekt mit der Lorentzkraft $\vec{F} = Q \cdot (\vec{v} \times \vec{B})$. Das Magnetfeld lenkt die im elektrischen Strom bewegten Teilchen (es sind meist Elektronen) ab. Wird statt des elektrischen Stromes ein Wärmestrom genommen, erzeugt durch einen Temperaturunterschied an den Rändern längs der Platte, tritt wie beim Halleffekt auch eine Querspannung auf und damit auch ein kleiner elektrischer Strom. Walther Nernst (1864–1941) entdeckte diesen zuletzt erwähnten und auch nach ihm benannten Effekt im Jahre 1887. Bei dem Halleffekt tritt aber noch zusätzlich neben der Hallspannung auch eine Temperaturdifferenz zwischen dem oberen und unteren Rand des Plättchens auf, ebenfalls verursacht durch das außen angelegte Magnetfeld. Dieser Effekt wurde 1887 von Andreas von Ettingshausen (1850–1932) entdeckt. Es sei bemerkt, dass die Ergebnisse bei den Untersuchungen zu den galvanomagnetischen und thermomagnetischen Wirkungen quantitativ zum Teil nur wenig, mit den experimentellen Beobachtungen übereinstimmen. Dasselbe gilt auch für die Drudeschen Theorien auf diesem Gebiet. Lorentz hat sich mit diesem Wissensgebiet wenig befasst. Diese Lücke füllte dann R. Gans mit seinen Arbeiten 1906 aus.[27]

Weil der Halleffekt bei einigen Metallen eine entgegengesetzte Polung der üblichen Hallspannung aufwies – man spricht hier von einem anomalen Halleffekt – beharrten Riecke und Drude auf der Existenz der positiven beweglichen elektrischen Teilchen. Lorentz dagegen wollte nur negative bewegliche elektrische Teilchen (Elektronen) zulassen. Im Falle der Deutung von Berührspannung sah Lorentz einen Verstoß gegen den zweiten Hauptsatz der Thermodynamik.[28] In seinem Vortrag vor dem Elektrotechnischem Verein 1904 liest man zu diesem Problem:[29]

> „Ich halte es für nicht ausgeschlossen, dass es ... gelingen wird, von dem [anomalen Halleffekt] Rechenschaft zu geben, ohne dass man zu freien positiven Elektronen Zuflucht zu nehmen braucht."

Mit den „Elektronenlöchern" oder „Defektelektronen" wurde später hier eine Lösung gefunden.

---

27 Siehe Seeliger 1921, S. 814. Gans, Richard: Annalen der Physik 20 (1906), S. 212.
28 Lorentz 1907, S. 125.
29 Lorentz 1905, S. 116.

Abbildung 2.9:
Paul Drude (1863–1906)
http://it.wikipedia.org/wiki/Paul_Drude

2.7 Paul Drude (1863–1906)

Paul Drude (1863–1906) setzte mit seiner Abhandlung „Zur Elektronentheorie der Metalle" zwei Jahre nach Riecke einen zweiten Markstein. Mit Rieckes Arbeit hat er viele Berührpunkte, aber auch manche grundsätzliche Verschiedenheit. So hat Riecke nur die Zusammenstöße der elektrischen Teilchen mit den Metallatomen in seiner Rechnung miteinbezogen, nicht aber die Zusammenstöße der Teilchen untereinander. Umgekehrt berücksichtigt Drude nun die Zusammenstöße zwischen elektrischen Teilchen. Seine Überlegungen schließen unmittelbar an die kinetische Gastheorie an. Bei Riecke ist die Dichte der beweglichen Ladungsträger klein gegenüber der Dichte der Metallatome, bei Drude dagegen ist die Dichte der Metallatome klein gegenüber der Dichte der Elektronen. Bei Drude geschieht der Wärmetransport wie bei Riecke durch die elektrischen Teilchen. Die Metallatome sind an feste Plätze gebunden und in ihren Schwingungen, die durch Zusammenstöße mit den elektrischen Teilchen entstehen berühren sich nicht gegenseitig.

Gehen wir näher auf die obige Abhandlung von Drude ein, sie ist geprägt von Drudes wissenschaftlichem Weg. Zwischen 1882 und 1887 studierte er, abgesehen von wenigen Gastsemestern an der Universität in Göttingen Mathematik und Physik bei Woldemar Voigt (1850–1919), und promovierte mit einem Thema aus der Kristallphysik. Dabei lernte er die mechanische Lichttheorie kennen, in der der Äther ein hochelastisches Medium ist, in dem sich die transversalen Lichtwellen ausbreiten. Als 1888 Heinrich Hertz (1857–1894) die elektromagnetischen Wellen entdeckt hatte, wurde auch Drude von der euphorischen Stimmung mitgerissen. In seinem Buch *„Physik des Äthers auf elektromagnetischer Grundlage"* (1894) wollte er die Maxwellsche Theorie angehenden und auch gestandenen Physikern zugänglich machen. Er setzte sich zum Ziel, die Optik in die Sprache der elektromagnetischen Lichttheorie zu überführen und dies tat er mit beachtlichem Erfolg. Um aber die optische Dispersion und Absorption an Metallen behandeln zu können, kam ihm die Entdeckung des Elektrons 1897 zu Hilfe. Er erweiterte die Grundlagen der Maxwellschen Theorie, und gelangte so schließlich zur Elektronentheorie der Metalle.[30]

In der Arbeit von 1900 will er die negativen einfachen Ladungsträger einer Elementarladung lieber als einfache Kerne (Kerne erster Gattung) bezeichnen. Er schließt nicht aus, dass diese sich zusammenballen, „polymerisieren" können, so z. B. zu negativen und positiven Doppelkernen ($+2e$, $-2e$) oder auch zu multiplen. Hierzu sei aber bemerkt, dass er von Doppel- und Mehrfachkernen wenig Gebrauch machte. Das Wort „Corpuskel" möchte er vermeiden, $+e$ und

---

30 Drude, Paul: Physik des Äthers auf elektromagnetischer Grundlage. Stuttgart 1894.

$-e$ sollten bei ihm am liebsten keine träge Masse im üblichen Sinne haben, sondern eine elektromagnetische Masse besitzen, entstanden bei Bewegung durch Selbstinduktion (Trägheit des Magnetfeldes). Ein einfacher Kern ist bei Drude nur eine singuläre Stelle im Äther, die eine Quelle oder Senke für elektrische Kraftlinien ist. Hier zeigt sich deutlich der Maxwell-Hertzsche Einfluß, und hier steht er im Gegensatz zu Riecke, der in der Weberschen Tradition verhaftet ist. Für die frei beweglichen einfachen Kerne akzeptiert Drude die kinetische Gastheorie und die Maxwell-Boltzmannsche Geschwindigkeitsverteilung. Bei seinen Rechnungen benutzte er allerdings nur mittlere Werte für Geschwindigkeit, Weglänge und Energie. Die freien positiven und negativen Kerne vereinigen sich nicht auf Grund ihrer relativ hohen kinetischen Energie. Positive Kerne sind wie auch bei Riecke an ponderable Metallatome gebunden und bilden die Kanalstrahlen. Für die kinetische Energie der beweglichen elektrischen Teilchen setzt Drude $\frac{1}{2}mu^2 = \alpha T$ und $\alpha = \frac{3}{2}k$. Dabei ist T die absolute Temperatur und k die Boltzmann-Konstante. Es werden dabei drei lineare Freiheitsgrade angenommen. Die Wärmeleitung berechnet Drude nach Ludwig Boltzmanns (1844–1906) „*Vorlesungen über Gastheorie*" (Leipzig: J. A. Barth 1896). Für das thermische Leitvermögen κ ergibt sich: $\kappa = \frac{1}{3}\alpha(u_1 l_1 N_1 + u_2 l_2 N_2 + ...)$. Dabei sind u die Geschwindigkeiten, l die Weglänge und N die Dichte der Teilchen.

Um die elektrische Leitfähigkeit σ zu finden, betrachtet Drude zunächst die Translationsgeschwindigkeit der Teilchen unter dem Einfluß der Spannung, definiert durch die Beweglichkeiten v, $v = \frac{l \cdot u}{4\alpha T}$ und erhält so

$$\sigma = \frac{1}{4\alpha T}\left((e \times 1)^2 N_1 \cdot l_1 \cdot u_1 + (e \times 2)^2 N_2 \cdot l_2 \cdot u_2 + ...\right)$$

Die elektrische Leitfähigkeit ist, wie das Gesetz von Rudolf Clausius verlangt, der absoluten Temperatur umgekehrt proportional. Die Quotientenbildung von k und σ führt zum Wiedemann-Franz-Lorenzschen Gesetz

$$\frac{\kappa}{\sigma} = \frac{4}{3}\left(\frac{\alpha}{e}\right)^2 \cdot T$$

unter der Voraussetzung, daß nur einfache Kerne vorhanden sind.

Max Planck (1858–1947) sah in der Ableitung dieses Gesetzes die bedeutendste Leistung von Drude.[31] Es gibt aber auch Ausnahmen bei diesem Gesetz, und Drude nennt hier insbesondere Wismut. Die Ursache sah er darin, dass die Dichte der beweglichen Ladungsträger im Gegensatz zu den Metallen von der Temperatur abhängig ist. Bei Wismut steigt mit zunehmender Temperatur die

---

31 Planck 1906, hier S. 621.

elektrische Leitfähigkeit, im Gegensatz zu den Metallen – eine Eigenschaft, die für Halbleiter charakteristisch ist.

Auf dem Gipfel seiner Karriere als Direktor des Physikalischen Instituts der Friedrich-Wilhelm-Universität Berlin, Mitglied der Berliner Akademie, Herausgeber der Annalen der Physik, schied Paul Drude 1906 freiwillig aus dem Leben, für Familie, Freunde und Bekannte überraschend und unerklärbar. Im „Moloch" Berlin mit seinen Verpflichtungen vielerlei Art fühlte er sich den Anforderungen nicht mehr gewachsen. In dem Buch „*Nachtgedanken eines klassischen Physikers*" versucht R. McCormmach den Gründen nachzuspüren.[32]

2.8 Ludwig Lorenz (1829–1891), dänischer Physiker

Nach Wiedemann-Franz ist der Quotient bei konstanter Temperatur annähernd konstant. Nach Lorenz ist dieser Quotient proportional zu T.[33] Bei tiefen Temperaturen verliert das Gesetz seine Gültigkeit. Max Planck sieht in der Ableitung des Wiedemann-Franzschen-Gesetzes die bedeutendste Leistung von Drude.[34]

Noch ein Ergebnis der Drudeschen Herleitungen ist zu beachten (Seeliger, 1921, S. 805). Bei Metallen ist das Wiedemann-Franzsche-Gesetz einigermaßen befolgt, bei Wismut z. B. weniger. Hier macht Drude die Annahme, das die Dichte der bewegten Ladungsträger von der Temperatur abhängig ist, im Gegensatz zu der Unabhängigkeit bei den Metallen. Die Annahme von temperaturunabhängiger Dichte muß bei Wismut fallengelassen werden. Denn bei Wismut steigt mit zunehmender Temperatur die elektrische Leitfähigkeit (Eigenschaft der Halbleiter) und weiter list man das bei solchen Ausnahmen auch die positiven Teilchen zur elektrischen Leitfähigkeit beitragen müssen. Mit der Vorstellung von einem Elektronengas konnte von vielen Erscheinungen der Elektrizitätsleitung Rechenschaft abgegeben werden. Doch Konsequenzen daraus führen zu groben Widersprüchen mit der Erfahrung. In erster Linie geht es hier um die Frage des Beitrages der Elektronen zur spezifischen Wärme des Metalls. Wir kommen noch darauf zurück.

Nach dem Gleichverteilungsgesetz der Maxwell-Boltzmannschen Statistik kommen jedem Freiheitsgrad der Elektronen die wie ein Gas behandelt werden der

---

[32] McCormmach 1984, S. 114–128.
[33] Wiedemann: Die Lehre vom Galvanismus und Elektromagnetismus..., 1861–1863.
[34] Siehe Abhandlungen und Vorträge, Bd. 3 (1958), S. 289. Der Quotient der thermischen und elektrischen Leitfähigkeit stimmt genau mit der von Joos 1945, S. 396, überein. Seeliger (1921) hat eine Zusammenstellung der Formeln von Riecke, Drude, Lorentz, Bohr und anderen auf S. 303 gegeben. Der empirisch gefundene Wert liegt zwischen dem von Drude und Riecke (2,51, 2,27). Empirscher Wert 2,4 wird angestrebt.

Energiebetrag $\frac{1}{2}k \cdot T$ zu. Wenn man annimmt, dass die Zahl der freien Elektronen mit der Zahl der Metallatomen über-einstimmt, muß der Energieinhalt eines Mols des betreffenden Metalls um $\frac{3}{4}R \cdot T$ vermehrt werden (R Gaskonstante). Für die Elektronen sind dabei drei Freiheitsgrade der angenommenen gradlinigen Bewegungen anzunehmen.

2.9 Hendrik Antoon Lorentz (1853–1928)

Spätestens im Jahre 1903 wandte sich Hendrik Antoon Lorentz (1853–1928) der Elektronentheorie der Metalle zu. Lorentz ist einer der Vollender der klassischen Physik. Im letzten Jahrzehnt des 19. Jahrhunderts hatte er die Elektronentheorie der Materie begründet. Damit begann die Erforschung und die Beschreibung der atomistischen Struktur der Materie. Lorentz war eine Synthese zwischen der Maxwellschen Theorie und der auf dem Festland vorhandenen älteren Elektrodynamik gelungen.[35]

Lorentz vervollkommt und verfeinert die Theorie von Drude, indem er statt der Mittelwerte die genaue statische Verteilung von Geschwindigkeit und Energie der Elektronen nach Maxwell zu Grunde legte, sie aber auf die Verhältnisse hin im Metall modifizierte. Lorentz nimmt im Gegensatz zu Riecke und Drude nur eine Art freier einfacher beweglicher Ladungsträger an, nämlich die negativen Teilchen, die Elektronen (Ladung und Masse identisch mit dem Kathodenstrahlteilchen). Der Schwarm der Elektronen wird in seiner Bewegung eingeschränkt, weil er an die Metallatome stößt. Mit Einbeziehung des elektrischen Feldes bei angelegter Spannung an den Leiter und des Temperaturgradienten beim Wärmetransport, bekommen die Lorentzschen Formeln Zusatzglieder. Für den Quotienten aus Wärmeleitfähigkeit κ und elektrischer Leitfähigkeit κ findet Lorentz:[36]

$$\frac{\kappa}{\sigma} = \frac{8}{9}\left(\frac{\alpha}{e}\right)^2 \cdot T$$

Überraschenderweise schneidet die Lorentzsche Formel beim Vergleich mit den empirischen Werten schlechter ab als die von Riecke und Drude. In seinem Vortrag vor dem elektrotechnischen Verein 1904 kommt Lorentz auch auf den

---

35 Lorentz' Arbeiten zur Metallelektronik: Lorentz (1892), [Arch. Nederl. (2) 10 (1905), S. 336. Proc. Acad. Amsterdam 7 (1905), S. 438, 585, 684.] - Lorentz: Ergebnisse und Probleme der Elektronentheorie (1905). - Lorentz: Jahrbuch der Radioaktivität und Elektronik 4 (1907), S. 125. - Lorentz: Theory of electrons (1909). - Lorentz: Anwendung der kinetischen Theorie auf die Elektronentheorie (1915).
36 Bemerkung: α hat die Bedeutung wie vorher bei Riecke und Drude.

Abbildung 2.10:
Hendrik Antoon Lorentz (1853–1928), 1916
portraitiert von Menso Kamerlingh Onnes (1860–1925)
(Wikipedia)

Halleffekt zu sprechen. Bei der Deutung benutzt er sein nach ihm benanntes Kraftgesetz $\vec{F} = Q \cdot (\vec{v} \times \vec{B})$. Es ist die Kraft, die auf eine mit der Geschwindigkeit v sich bewegenden Ladung Q im Magnetfeld $\vec{B}$ ausgeübt wird (Vektorprodukt). Merkwürdigerweise hat Lorentz, wie schon erwähnt, zu den anderen galvanomagnetischen und thermomagnetischen Effekten keine ausgearbeitete Theorie gegeben.

Weil neben der normalen Halbspannung beim negativen Halleffekt, die durch Elektronen hervorgerufen wird, auch eine Halbspannung mit entgegengesetzter Polung auftreten kann, sahen sich Riecke und Drude in ihrer Annahme auch von positiven Teilchen bestätigt. Dies lehnte Lorentz, wie schon erwähnt, entschieden ab, und er glaubte, dass sich eine Lösung für den anomalen Hall-Effekt finden werde – die Elektronenlöcher und Defektelektronen taten dieses.

2.10 Niels Bohr (1885–1962)

Wenig bekannt ist, dass Niels Bohr (1885–1962) seine Dissertation über die Elektronentheorie der Metalle geschrieben hat.[37] Niels Bohr betrachtet darin kritisch die Annahmen seiner Vorgänger sowohl in physikalischer als auch in statistischer Hinsicht. Er berücksichtigt die Zusammenstöße sowohl zwischen Elektronen untereinander als auch solche mit den Metallatomen. Aber statt der freien Bahnen zwischen zwei Zusammenstößen sieht er wegen der dichten Packung der Metallatome die Elektronen großen Teils in der Wirkungssphäre der Metallatome verbleibend, und deren Kräfte wirken auf die Bewegung der Elektronen ein. Es sind Zentralkräfte und die Verteilungsfunktionen werden komplizierter. Für die elektrische Leitfähigkeit und die thermische Leitfähigkeit sind seine Formeln komplizierter. Nimmt man wieder vollkommen elastische Zusammenstöße an, gehen Bohrs Formeln in diejenigen von H. A. Lorentz über.

In seiner Dissertation diskutierte Bohr auch die para- und diamagnetischen Eigenschaften der Körper. Dabei machte er auf die Arbeiten von Paul Langevin (1872–1946) aufmerksam. Dieser beruft sich neben anderem aber auch auf die Ampère-Webersche Theorie, genauer auf die Spekulationen zur Erzeugung des Para- und Diamagnetismus von W. Weber.[38]

Bohrs Dissertation erntet hohes Lob bei skandinavischen Wissenschaftlern, aber das Lesen der zunächst in Dänisch verfassten Schrift stößt bei ausländi-

---

[37] Bohr, Niels: Studier over Metallernes Elektrontheori. Copenhagen 1911. Es gibt auch eine englische Übersetzung. Ein längeres Referat darüber gibt Seeliger 1921, bes. S. 795–798. - Ebenfalls Sutter, P.: Die Elektronentheorie der Metalle. Bern 1920. Siehe auch Röseberg 1985, S. 43 ff.

[38] Langevin: Magnétism et Théorie des Electrons (1905), hier S. 73 und 96.

Abbildung 2.11:
Niels Bohr (1885–1962)
http://da.wikipedia.org/wiki/Fil:Niels_Bohr.jpg

schen Wissenschaftlern auf Sprachprobleme. 1911 ging Bohr nach Cambridge in England, das damals eine magische Anziehungskraft hatte. Dort legte er seine in Englisch verfasste Dissertation Joseph John Thomson (1856–1940) vor, dem Entdecker des Elektrons. Für Niels Bohr wurde es eine Enttäuschung. Thomson zeigte nämlich aus Zeitmangel wenig Interesse. Als Bohr im gleichen Jahr den aus Neuseeland stammenden Ernest Rutherford (1871–1937) kennenlernte und die beiden Sympathie für einander empfanden, wechselte Bohr 1912 nach Manchester über. Dort hatte Rutherford seit 1907 den Lehrstuhl für Physik. Beim Analysieren der Streuung von Alphastrahlen an Atomen hatte Rutherford 1910 mit seinem Assistenten Hans Geiger (1882–1945) den positiv geladenen

Atomkern entdeckt. Für den jungen Bohr war dies der entscheidende Schritt gewesen hin zu einem Weg der ihn in der Atomphysik zur Weltgeltung führte.

2.11 Krise der klassischen Elektronentheorie der Metalle

Der schönste Erfolg der klassischen Elektronentheorie der Metalle war die Herleitung des Wiedemann-Franz-Lorenzschen Gesetzes durch Drude.

$$\frac{\kappa}{\sigma} = \frac{4}{3}\left(\frac{\alpha}{e}\right)^2 \cdot T \quad,$$

Riecke fand vier Jahre später

$$\frac{\kappa}{\sigma} = \frac{3}{2}\frac{\alpha^2}{e^2}\left(1 + \frac{2}{3}\cdot \delta \cdot t\right) \quad.$$

Lorentz fand 1905 die Formel

$$\frac{\kappa}{\sigma} = \frac{8}{9}\left(\frac{\alpha}{e}\right)^2 \cdot T \quad.$$

Es schien, als ob damit eine tragfähige Theorie der Metallelektronik gefunden worden sei – ein Irrtum. Peter Debye (1884–1966) entwickelte nach dem ersten Jahrzehnt des 20. Jahrhunderts seine Gittertheorie der Festkörper. Die Gitteratome, die durch Wechselwirkungskräfte festgehalten werden, schwingen um ihre Ruhelage (Debye 1914). Die kinetische Energie der schwingenden Gitteratome liefert für die spezifische Wärme der Metalle $\frac{3}{2}R$ (R = Gaskonstante). Hinzu kommen aber durch die potentielle Energie nochmals $\frac{3}{2}R$, insgesamt also $\frac{6}{2}R$. Nach den Gesetzen der Thermodynamik muss das Elektronengas nochmals $\frac{3}{2}R$ zur spezifischen Wärme beitragen. $\frac{9}{2}R$ widerspricht aber allen empirischen Werten, denn die 1819 von Pierre Louis Dulong (1785–1838) und Alexis Thérèse Petit (1791–1820) gefundene Regel über die Atomwärme besagt, dass die spezifische Wärme multipliziert mit dem Atomgewicht immer etwa 6 ist.[39] Dies ist aber gleichbedeutend damit, dass die spezifische Wärme nur $\frac{6}{2}R$ sein darf.[40]

Zu den mannigfachen theoretischen Versuchen, die Schwierigkeiten der älteren gaskinetischen Theorie zu beheben, zählen auch die sogenannten Gittertheorien. Typischer Vertreter einer solchen Theorie ist Frederick A. Lindemann (1886–1957), bekannt als Widersacher bei der Berufung von Arnold

---

39 Siehe Rosenberger 1887–1890, (1965), Teil 3, S. 221 und Tomaschek: Grimsehls Lehrbuch der Physik, 1. Bd., 1944, S. 443.
40 Joos 1945, S. 396.

Sommerfeld nach München und als Berater und Befürworter der Bombardierung deutscher Städte im Zweiten Weltkrieg. Nach ihm bilden nicht nur die Metallatome, sondern auch die Elektronen kein statistisches Ensemble, sondern sind auch in einem Gitter regelmäßig angeordnet. Die Elektronen sind also nicht mehr freibeweglich. Lindemann war ein entschiedener Gegner der Auffassungen von Drude und Lorentz.

Mit allen möglichen und unmöglichen Annahmen suchten die Theoretiker der Metallelektronik nach Auswegen aus diesem Dilemma.[41] Lorentz will die Elektronendichte gegenüber der Metallatomdichte äußerst gering machen.[42] Riecke glaubt, dass die potentielle Energie der schwingenden Metallatome reduziert werden könnte.[43] Joseph John Thomson, der sich drei Jahre nach seiner Entdeckung des Elektrons der metallischen Leitung zuwandte, wollte die Elektronen nicht als ein Gas sehen, das den Raum zwischen den Gitteratomen ausfüllt.[44] Viele Physiker wandten sich wegen dieses ungelösten Problems von der Elektronentheorie der Metalle ab. Dies ist umso erstaunlicher, da bereits 1914 experimentelle, schlagkräftige Beweise für die Lorentzschen freibeweglichen Elektronen in einem Metall gefunden worden waren.[45]

2.12 Stewart-Tolman-Versuch – Richard Chace Tolman (1881–1948) und Thomas Dale Stewart (1890–1958)

Die Grundidee für das Stewart-Tolman-Experiment[46] hatte – wie vorher schon erwähnt – James Clerk Maxwell 1873 ausgesprochen. Tolman und Stewart versetzten eine flache Spule aus vielen Windungen Kupferdrahtes um eine zur Windungsfläche senkrecht stehende Achse in schneller Rotation. Die zwischen dem Gitter der Metallatome befindlichen Elektronen nahmen nach einer genügenden Zeit die Umfangsgeschwindigkeit der Spule an. Die Spule war an ihren Enden über Schleifringe mit einem ballistischen Galvanometer verbunden. Bei plötzlichem Anhalten der Spule entstand durch die gesamte kinetische Energie der Elektronen ein Stromstoß, den das Galvanometer anzeigte. Aus der so gemessenen Elektrizitätsmenge, der Umdrehungsgeschwindigkeit, der Länge des Drahtes und seinem Widerstand konnte die Größe der Ladung eines einzelnen

---

41 Benedicks, Carl (1916), hier 359.
42 Lorentz 1913, S. 216.
43 Riecke 1909, S. 516. Bohr 1911.
44 Thomson, Joseph John 1908, S. 48.
45 Lorentz 1905, 1907, 1909 und 1915. Seeliger 1921, S. 863.
46 Tolman und Stewart (1914), S. 466–485 und (1916), S. 97–116.

Abbildung 2.12:
Richard Chace Tolman and Albert Einstein, 1932
http://en.wikipedia.org/wiki/File:Tolman_&_Einstein.jpg

beweglichen Elektrizitätsteilchens errechnet werden. Spätere Versuche mit einer verbesserten Apparatur ergaben einen $\frac{e}{m}$-Wert, der mit dem in anderen Experimenten gefundenen Wert genügend übereinstimmte.

Merkwürdigerweise beachtete man das Tolman-Stewartsche Experiment lange Zeit nicht. Auch im 1919 erschienen zweiten Band des *„Lehrbuchs der Physik"* von E. Riecke findet man keine Notiz darüber.

2.13 Arnold Sommerfeld (1868–1951)

Einen endgültigen Durchbruch der Elektronentheorie der Metalle, von vielen Physikern schon zu schnell ad Acta gelegt, gelang Ende der zwanziger Jahre des 20. Jahrhunderts Arnold Sommerfeld. Zuvor war die Quantentheorie weiterentwickelt worden und von Erwin Schrödinger (1887–1961) die Wellenmechanik geschaffen.

Walther Nernst (1864–1942) hatte sein Wärmetheorem aufgestellt und den Abfall der spezifischen Wärme fester Körper mit sinkender Temperatur experimentell bestätigt. Sollten Gase sich nicht ähnlich verhalten? Versuche zu einer „Gasentartungstheorie" gab es mehrere.[47] Wilhelm Wien (1864–1928) war mit seinem quantentheoretischen Ansatz in seiner Abhandlung „Zur Theorie der elektrischen Leitung in Metallen" 1913 am konsequentesten. Er entkoppelte das Elektronengas von der Wärmebewegung; die Geschwindigkeit der Elektronen wurde so unabhängig von der Temperatur – ein Gedanke, der sich in der Sommerfeldschen Theorie wiederfindet.

Der entscheidende Schritt von Sommerfeld war, dass er die Statistik von Enrico Fermi (1901–1954) und Paul Adrian Maurice Dirac (1902–1984) heranzog. Zuvor hatte der englische Physiker Satyendra Nath Bose (1894–1974) eine ganz neuartige Statistik für die Photonen entwickelt, die dann von Albert Einstein auf materielle Teilchen erweitert wurde (Bose-Einstein-Statistik). Zum Unterschied gegenüber der Maxwell-Boltzmannschen-Statistik (manchmal auch Statistik des normalen Menschenverstandes genannt) wird die individuelle Unterscheidung der Teilchen aufgegeben, die wahrscheinlichkeitstheoretisch erfasst werden. Der Bose-Einstein-Statistik und auch der Fermi-Dirac-Statistik liegen Besetzungszahlen für die verschiedenen Quantenzustände zu Grunde. In der klassischen Statistik konnten z. B. beliebig kleine Werte der kinetischen Energie der Teilchen angenommen werden. Bei der neuen Statistik ist die Größe der „Elementarzelle" quantenmäßig durch das Plancksche Wirkungsquantum bzw. durch andere quantenmäßige Größen beschränkt.

Enrico Fermi verknüpfte die neue Statistik mit dem Pauli-Prinzip (scherzweise das Wohnungsamt der Elektronen genannt). Diese Prinzip besagt, dass in einem einzelnen Atom jeder vollständig definierte Quantenzustand höchstens von <u>einem</u> Elektron eingenommen werden kann. Die Theorie des Periodensystems der chemischen Elemente konnte so ihren Abschluß finden. Das Pauli-Prinzip gilt auch für die in einem zusammengesetzten Molekül vereinten Elektronen. Fermi tat nun den kühnen Schritt, dieses Prinzip auf alle Moleküle eines ausgedehnten Gases zu übertragen. Aus der Fermi-Statistik folgt ein ganz bestimmtes Gesetz für die „Entartung" einatomiger Gase und auch die Existenz einer Null-Punkts-Energie, d. h., am absoluten Nullpunkt hat ein Gas noch (wenn auch sehr kleine) kinetische Energie.

Fermi gibt auch ein Entartungskriterium an. Für das Elektronengas zeigt sich, dass es schon bei Laboratoriumstemperaturen vollkommen entartet ist. Begünstigt wird die Entartung durch die hohe Dichte der Elektronen (es wird ein Elektron pro Metallatom angenommen) und die kleine Masse des Elek-

---

47 Siehe Seeliger 1921.

Abbildung 2.13:
Arnold Sommerfeld (1868–1951) – 1897
http://www.lrz.de/~Sommerfeld/Bilder/as97_01.gif (Wikimedia Commons)

Abbildung 2.14:
Enrico Fermi (1901–1954), 1943/49
(Wikipedia)

trons.⁴⁸ Für kleine Dichten führt die Fermi-Statistik zu der Maxwell-Boltzmannschen-Statistik.

Sommerfeld wurde zu seiner Abhandlung durch die Arbeit von Wolfgang Pauli „Über Gasentartung und Paramagnetismus" angeregt, in der mit Erfolg die Fermi-Statistik benutzt wurde.⁴⁹ Bei seinen Rechnungen erhält Sommer-

---

48 Sommerfeld, A.: Zur Elektronentheorie der Metalle auf Grund der Fermischen Statistik. In: Die Naturwissenschaften 15 (1927),No. 41, S. 825–832. - Zur Elektronentheorie der Metalle auf Grund der Fermi'schen Statistik. 1. Allgemeines, Strömungs- und Austrittsvorgänge. In: Zeitschrift für Physik Bd. 47 (1928) S. 1–32. - Sommerfeld, Bethe: Elektronentheorie der Metalle. In: Geiger/Scheel: Handbuch der Physik. 2. Aufl., Bd. 24, 2. Teil, Berlin 1933, S. 443–622; die ersten Kapitel hat Sommerfeld verfasst. - Wien, Harms: Handbuch der Experimentalphysik, Bd. 11, 2. Teil, S. 272.

49 Pauli, Wolfgang: Über Gasentartung und Paramagnetismus. In: Zeitschrift für Physik 41 (1926), S. 81–102.

Abbildung 2.15:
Paul Adrian Maurice Dirac (1902–1984)
http://www.learn-math.info/history/photos/Dirac_3.jpeg

feld in erster Näherung, dass die innere Energie des Elektronengases von der Temperatur unabhängig ist. Da man die spezifische Wärme durch Differenzierung nach der Temperatur erhält, ergibt sich für diese der Wert Null ($c_v = 0$). Damit ist das größte Hindernis beseitigt, und die spezifische Wärme für das Metall nicht mehr $\frac{9}{2}R$, sondern wie verlangt nur $\frac{6}{2}R$. In der Herleitung der elektrischen und thermischen Leitfähigkeit lehnt sich Sommerfeld an die Methode von H. A. Lorentz an, benutzt aber auch die Elektronenwellen nach de Broglie und die Wellenmechanik von Schrödinger. Bei der thermischen Leitfähigkeit muß Sommerfeld die zweite Näherung für die Energie heranziehen. Er findet für die elektrische Leitfähigkeit:⁵⁰

$$\sigma = \frac{4\pi}{3} \cdot \frac{e^2 l}{h} \cdot \left(\frac{3n}{4\pi}\right)^{\frac{2}{3}}$$

für die Wärmeleitfähigkeit:

$$\kappa = \frac{4\pi^3}{9} \cdot \frac{lk^2 \cdot T}{h} \cdot \left(\frac{3n}{4\pi}\right)^{\frac{2}{3}}$$

Dabei ist e das elektrische Elementarquantum, h das Plancksche Wirkungsquantum, l die Weglänge, k die Boltzmann-Konstante, n die Dichte der Elektronen und T die absolute Temperatur. Die Quotientenbildung ergibt:

$$\frac{\kappa}{\sigma} = \frac{\pi^2}{3}\left(\frac{k}{e}\right)^2 \cdot T$$

Das ist aber das universell gültige Wiedemann-Franz-Lorenzsche-Gesetz. Setzt man bei der Drudeschen Formel $\frac{4}{3}(\frac{\alpha}{e})^2 \cdot T$ für α gleich $\frac{3}{2}k_B$ ein, so erhält man $3(\frac{k}{e})^2 \cdot T$ und bei der Lorentzschen Formel $\frac{8}{9}(\frac{\alpha}{e})^2 \cdot T$ erhält man $2(\frac{k}{e})^2 \cdot T$.

Bei Sommerfeld ist der Zahlenkoeffizient etwa 3,3. Dieser Faktor schließt sich noch näher an die Beobachtungen an, als der Drudesche und Lorentzsche Faktor. Es sei aber darauf hingewiesen, dass nach der Art der Sommerfeldschen Ableitung die Formel für tiefe Temperaturen keine Gültigkeit hat. Auf die verbesserte Behandlung der Emission von Elektronen bei erhitzten Metallen (Glühelektronen, Richardsoneffekt) Kontaktelektrizität und anderes durch Sommerfeld soll nicht eingegangen werden.

Sommerfelds Arbeit war nur ein erster Schritt. Eine Vertiefung und Verfeinerung der Elektronentheorie der Metalle erfolgte kurz darauf durch den aus der Schweiz stammenden Physiker Felix Bloch (1905–1983), erster Doktorand bei Heisenberg in Leipzig,⁵¹ und durch die Schüler und Mitarbeiter von Sommerfeld, Rudolf Ernst Peierls (1907–1995) und Hans Bethe (1906–2005), der mit Sommerfeld den Artikel über Elektronentheorie der Metalle in dem Handbuch der Physik (2. Aufl., Berlin 1933) verfasste.⁵² Als 60jähriger hatte Sommerfeld

---

50 Sommerfeld (1927), S. 827.
51 Bloch verfasste grundlegende Arbeiten in der Festkörperphysik. In den USA machte er eine steile Karriere.
52 Literatur zu Sommerfeld: Eckert et al.: Geheimrat Sommerfeld. München 1984. Bemerkung: Die Elektronentheorie der Metalle ist darin kaum behandelt. In seiner Biografie über Sommerfeld (in Meÿenn 1997, Bd. 2, S. 208) gibt er allerdings einen kurzen Überblick über die Sommerfeldsche Arbeit von 1927 und über die Bedeutung für die Festkörperphysik, vgl. Eckert 1990.

mit seiner *Elektronentheorie der Metalle* (1927 und 1933) nochmals einen ähnlichen Erfolg wie mit seinem Buch „*Atombau und Spektrallinien*" (1919), das damals als „Bibel der Atomphysik" bezeichnet wurde.[53]

Die Hypothese der freien Elektronen mußte durch ein Modell ersetzt werden, das der Wirklichkeit näher kam und das Gitter der Metallatome mit einbezog. Johannes Stark (1874–1957) hatte schon in seiner „Atomdynamik" (1912) von einer regelmäßigen Anordnung der Metallatome gesprochen.[54] Quantitativ hatte sich Peter Debye, wie schon erwähnt, mit der Gittertheorie befasst und den Festkörper als Ganzes betrachtet.[55] Die Gitteratome werden durch Wechselwirkungskräfte festgehalten und schwingen um ihren Ruheplatz.

Um das Verhalten von Elektronen in einem Metall oder Kristall erklären zu können, beschreibt Felix Bloch die Bewegung der fast freien Leitungselektronen im Potentialfeld der Restatome. Das elektrische Feld wird durch die Periodizität und die Symmetrie des Metalls oder Kristalls bestimmt. Die Metalle besitzen meist ein kubisches Gitter. Das Gitter von Kupfer ist zum Beispiel kubisch flächenzentriert. Bloch untersucht mit Hilfe der de Broglie-Welle die Eigenwerte eines Elektrons. Wegen der vielen gekoppelten Nachbarn eines Gitteratoms findet eine Aufspaltung des Energieniveaus in ein ganzes Energieband statt. Das Energiespektrum besteht aus mehreren Bändern, zwischen den einzelnen Bändern, so findet Bloch, liegen „verbotene Zonen", Energiezustände, die von Elektronen nie angenommen werden können. Absorbiert ein Elektron einen Energiequant, kann es in ein höheres Band springen, wenn dort ein „Platz" frei ist. Ein Niveau innerhalb eines Bandes kann nur mit höchstens einem Elektron besetzt werden, – es gilt das Pauli-Prinzip.

Bei den Metallen ist das höchste Band nur halb besetzt oder bei Überlappung mit einem anderen höheren Band teilweise. Durch eine angelegte Spannung werden die Elektronen beschleunigt, springen dabei auf ein höheres Niveau, und geben dann aber Energie an die Gitteratome ab; das heißt, der Leiter erwärmt sich.

Alan Herries Wilson (1906–1995) konnte 1933 mit dem Bändermodell die Unterschiede zwischen Metallen, Halbleitern und Isolatoren deuten. Wenn das Valenzband ganz aufgefüllt ist und zwischen dem nächst höheren leeren Band eine breite verbotene Zone existiert, so kann der Festkörper den elektrischen Strom

---

53 Siehe dazu Hermann 1972, S. 351–355.
54 Stark, J.: Prinzipien der Atomdynamik, 3 Bde. Leipzig 1910–1914. Hermann 1972, S. 360 und Jahrbuch der Radioaktivität und Elektronik 9 (1912), S. 181 und Seeliger 1921, S. 859.
55 Siehe dazu Annalen der Physik 4, Bd. 33 (1910), S. 441–489, ebenso S. 1427–1434. - Annalen der Physik Bd. 39 (1912), S. 798. - Bd. 43 (1914), S. 49. - Physikalische Zeitschrift 13 (1912), S. 97. Siehe auch Meÿenn: Debye und sein Einfluß auf die Entwicklung der Atom- und Molekülphysik. In: Berlinische Lebensbilder, Bd. 1 (1987).

Abbildung 2.16:
Felix Bloch (1905–1983) – 1961
(Stanford University)

nicht leiten, man hat dann einen Isolator. Ist die verbotene Zone schmal, so stellt man bei Zimmertemperatur eine geringe Leitfähigkeit fest. Einige Elektronen haben die verbotene Zone schon übersprungen. Elemente oder Verbindungen, deren Kristallverband eine solche Energiebandkonfiguration haben, nennt man Halbleiter. Bei Erwärmung nehmen weitere Elektronen thermisch Energie auf, überspringen wieder die verbotene Zone und gelangen in das Leitungsband. Das heißt, die Leitfähigkeit nimmt zu – typisch für Halbleiter. Zu den Halbleiter gehören z. B. Bor, Schwefel, Selen, Tellur, Germanium und Silizium. Die beiden letzt genannten Elemente sind für die Halbleitertechnik besonders wichtig.

Der Schüler und Mitarbeiter von Sommerfeld, Rudolf Peierls (1907–1995) und Werner Heisenberg (1901–1976), konnten das Rätsel des anomalen Halleffektes (positive Hallspannung) lösen. Man beobachtet ihn bei Zink, Cadmium, Iridium und Halbleitern. Sie führten 1929 und 1931 die Elektronenlöcher oder Defektelektronen ein, die positiven Elektronen äquivalent sind.[56]

2.14 Literatur

AMPÈRE, ANDRÉ MARIE: Mémoire sur la théorie mathématique des phénomènes électro-dynamiques uniquement déduite de l'expérience. (1823) In: *Mémoires de l'Académie Royale des Sciences de l'Institut de France* 6 (1827), p. 175–387, hier S. 299.

Assis, Andre Koch Torres: Weber's Electrodynamics. Doordrecht (Niederlande): Kluwer Academic Publishers 1994.

Benedicks, Carl: Beiträge zur Kenntnis der Elektrizitätsleitung in Metallen und Legierungen. In: Jahrbuch der Radioaktivität und Elektronik 13 (1916), Heft 4, S. 351–395, hier 359.

Bloch, Felix: Über die Quantenmechanik der Elektronen im Kristallgitter. In: Zeitschrift für Physik 52 (1928), S. 555–600.

Bohr, Niels: Studier over Metallernes Elektrontheori. Copenhagen (Dissertation) 1911. Auch in: Niels Bohr Collected Works, Vol. 1. Amsterdam, New York: Elsevier 1972.

Debye, Peter: Zur Theorie der spezifischen Wärmen. In: Annalen der Physik (4) 39 (1912), S. 798–939.

Debye, Peter: Interferenz von Röntgenstrahlen und Wärmebewegung. In: Annalen der Physik (4) 43 (1914), S. 49–95.

---

[56] Heisenberg (1931), hier S. 901. Peierls, Rudolf Ernst (britischer Physiker deutscher Herkunft): Zur Theorie der galvanomagnetischen Effekte. In: Zeitschrift für Physik 53, Berlin: J. Springer 1929, S. 255–266.

Drude, Paul: *Physik des Äthers auf elektromagnetischer Grundlage*. Stuttgart: Verlag F. Enke 1894.

Drude, Paul: Zur Ionentheorie der Metalle. In: Physikalische Zeitschrift 1 (1900), S. 161–165.

Drude, Paul: Zur Elektronentheorie der Metalle. Teil I. In: Annalen der Physik 306, (4) 1 (1900), S. 566–613.
Zur Elektronentheorie der Metalle. Teil II. In: Annalen der Physik 308, (4) 3 (1900), S. 369–402.
Zur Elektronentheorie der Metalle. Teil III Berichtigung. In: Annalen der Physik 312, (4) 7 (1902), S. 687–692.

Eckert, Michael; Pricha, Willibald; Schubert, Helmut und Gisela Torkar: Geheimrat Sommerfeld. Eine Dokumentation aus seinem Nachlaß. München: Deutsches Museum 1984.

Eckert, Michael: Das „freie Elektronengas" – Vorquantenmechanische Theorien über die elektrischen Eigenschaften der Metalle. In: Deutsches Museum – Wissenschaftliches Jahrbuch 1989. München: R. Oldenbourg 1989, S. 57–91.

Eckert, Michael: Sommerfeld und der Anfang der Festkörperphysik. In: Deutsches Museum, Wissenschaftliches Jahrbuch, München 1990, S. 33–71.

Eckert, Michael: Die Atomphysiker. Eine Geschichte der theoretischen Physik am Beispiel der Sommerfeldschule. Braunschweig: Vieweg 1993.

Faraday, Michael: Historical Sketch of Electro-Magnetism. In: *Annals of Philosophy*, New Series, Vol. 2 (1821) und Vol. 3 (1822).

Faraday, Michael: *Experimental Researches in Electricity. 3 Vol.* London: R. Taylor & W. Francis 1839–1855. Deutsche Übersetzung Salomon Kalischer: *M. Faraday Experimental-Untersuchungen über Elektricität, 3 Bände*. Berlin: Julius Springer 1889–1891.

Fechner, Gustav Theodor: Über die Verknüpfung der Faraday'schen Inductions-Erscheinungen mit den Ampereschen electro-dynamischen Erscheinungen. In: Annalen der Physik und Chemie, hg. von Johann Christian Poggendorff (2) 64 (1845), S. 337–345.

Gans, Richard: Zur Elektronenbewegung in Metallen. Annalen der Physik 20 (1906), S. 212.

Geiger, Hans und Karl Scheel (Hg.): Handbuch der Physik. 26 Bände. Berlin: Springer (2. Auf|lage) 1926–1933.

Giese, Wilhelm: Grundzüge einer einheitlichen Theorie der Electricitätsleitung. In: Annalen der Physik und Chemie N. F., hrsg. von Gustav Wiedemann, (3) 37 (1889), S. 576–609.

Grahn, Holger T. und Dieter Hoffmann (Hg.): Drude, Paul: Zur Elektronentheorie der Metalle. Frankfurt am Main: Harri Deutsch (Ostwalds Klassiker der exakten Wissenschaften; 298) 2006. Darin Dieter Hoffmann: Leben und Werk, S. 7–22, Schriftenverzeichnis, S. 37–51.

Heisenberg, Werner: Zum Paulischen Ausschließungsprinzip. In: Annalen der Physik (5) 402 (1931), S. 888–904.

Hermann, Armin: Lexikon Geschichte der Physik A–Z. Köln: Aulis Verlag Deunbner & Co 1972.

Hermann, Armin (Hg.): Lexikon Geschichte der Physik. Köln: Aulis Verlag 2007.

Hoddeson, Lillian H. and Gordon Baym: The development of the quantum-mechanical electron theory of metals: 1928–1933. In: Proceedings of the Royal Society of London. A. Mathematical and Physical Sciences 371 (1980), S. 8–23.

Hoddeson, Lillian; Braun, Ernest; Teichmann, Jürgen and Spencer Weart: Out of the Crystal Maze: A History of Solid State Physics, 1900–1960. New York, Oxford: Oxford University Press 1992.

Hoddeson, Lillian; Baym, Gordon and Michael Eckert: The Development of the Quantum Mechanical Electron Theory of Metals, 1926–1933. In: Hoddeson et al.: Crystal Maze 1992, S. 88–181.

Hoffmann, Dieter; Laitko, Hubert und Staffan Müller-Wille (Hg.): Lexikon der bedeutenden Naturwissenschaftler. Heidelberg, Berlin: Spektrum Akademischer Verlag, München: Elsevier, Band 1 (A–E) 2003, Band 2 (F–Mei) 2004, Band 3 (Men–Z) 2004.

Hoffmann, Dieter: Paul Drude (1863–1906). In: Annalen der Physik (8) 15 (2006), S. 449–460.

Joos, Georg: Lehrbuch der theoretischen Physik. Leipzig: Akademische Verlagsgesellschaft (Becker & Erler) (7. Auflage) 1945.

Kant, Horst: Die Entstehung und Entwicklung der Festkörperphysik. In: Schreier, Wolfgang (Hg.): Geschichte der Physik – ein Abriss. Berlin: VEB Deutscher Verlag der Wissenschaften 1988, S. 363–373. Berlin, Diepholz: Verlag für Geschichte der Naturwissenschaft und Technik (2. Auflage) 1991, 2002.

Kohlrausch, Friedrich: Leitfaden der praktischen Physik. Leipzig: Teubner 1870.

Kohlrausch, Friedrich: Über Thermoelektricität, Wärme und Electricitätsleitung. Annalen der der Physik und Chemie, hg. von Johann Christian Poggendorff, (2) 156 (1875), S. 601–618.

Kohlrausch, Friedrich (Hg.): Fünf Abhandlungen über absolute elektrische Strom- und Widerstandsmessung. Leipzig: W. Engelmann (Ostwalds Klassiker der exakten Wissenschaften; 142) 1904.

Langevin, Paul: Magnétism et théorie des electrons. In: Annales de Chemie et de Physique (8. Ser.), T. 5 (1905), S. 70–127.

Lindemann, Frederick Alexander: Note on The Theory of the Metallic State. In: Philosophical Magazine and Journal of Science (6) 29 (1915), S. 127–140.

Lorentz, Hendrik Antoon: La Théorie electromagnétique de Maxwell et son application aux corps mouvants. In: Archives néerlandaises des sciences exactes et naturelles 25 (1892), S. 363–552.

Lorentz, Hendrik Antoon: Ergebnisse und Probleme der Elektronentheorie. Vortrag gehalten am 20. Dezember 1904 im Elektronischen Verein zu Berlin. Berlin: Springer 1905. Heerhugowaard 1905. (Auch in: Coll. Pap., Vol. 8. The Hague 1935, S. 76–124.)

Lorentz, Hendrik Antoon: Le movements des électrons dans les métaux. In: Archives néerlandaises des sciences exactes et naturelles 10 (1905), S. 336. (Auch in: Coll. Pap., Vol. 3. The Hague: Martinus Nijhoff 1935–1939, S. 180–214.)

Lorentz, Hendrik Antoon: On positive and negative electrons. Vortrag gehalten vor der American Philosophical Society vom 17. April 1906. Übersetzt ins Deutsche von Max Iklé. In: Jahrbuch der Radioaktivität und Elektronik 4 (1907). Hrsg. von Johannes Stark. Leipzig: S. Hirzel 1907, S. 125–131.

Lorentz, Hendrik Antoon: The theory of electrons. Leipzig, Berlin: B. G. Teubner 1909.

Lorentz, Hendrik Antoon: Theorie der magneto-optischen Phänomene (1909). In: Encyklopädie der Mathematischen Wissenschaften, 5. Bd. (Physik), 3. Teil, Leipzig 1909–1926, S. 256.

Lorentz, Hendrik Antoon: Anwendung der kinetischen Theorie auf die Elektronentheorie. Wolfskehl-Vortrag, April 1913. In: Mathematische Vorlesungen an der Universität Göttingen 6 (1914), S. 169. (Auch in: Coll. Pap., Vol. 8. The Hague: Martinus Nijhoff 1935–1939, S. 214–243.)

Lorentz, Hendrik Antoon: Weiterbildung der Maxwellschen Theorie. In: Encyklopädie der Mathematischen Wissenschaften mit Einschluss ihrer Anwendungen, 5. Band, 2. Teil (1921). Leipzig: B. G. Teubner 1904–1922, S. 63–144, 145–280.

Lorentz, Hendrik Antoon: Collected Papers, 9 Vol. The Hague: Martinus Nijhoff 1934–1939, Vol. 8 (1935). Study ed. by G. L. de Haas-Lorentz (tr. 1957).

Maxwell, James Clerk: Treatise on Electricity and Magnetism, Vol. 1 und Vol. 2 Oxford 1873, (Third Edition) 1904. Autorisierte deutsche Übersetzung von Dr. B. Weinstein: Lehrbuch der Electricität und des Magnetismus, Bd. 1 und Bd. 2. Berlin: Julius Springer 1883.

McCormmach, Russell: *„Nachtgedanken eines klassischen Physikers."* Frankfurt am Main: Insel Verlag 1984.

Meÿenn, Karl von (Hg.): Die großen Physiker, Band 1 und 2. München: C. H. Beck 1997.

Meÿenn, Karl von: Debye und sein Einfluß auf die Entwicklung auf die Atom- und Molekülphysik. In: Berlinische Lebensbilder, Band 1, Naturwissenschaftler. Hg. von Wilhelm Treue und Gerhard Hildebrandt. Berlin: Colloquium Verlag 1987.

Neumann, Carl: *„Explicare tentatur quomodo fiat ut lucis polarisationis per vires electricas vel magneticas declinetur."* Halis saxonum 1858. Deutsch: Die magnetische Drehung der Polarisationsebene des Lichtes. Halle 1863.

Neumann, Franz: Über ein allgemeines Princip der mathematischen Theorie inducirter elektrischer Ströme (1847), herausgegeben von C. Neumann. Leipzig (Ostwald's Klassiker der Exakten Naturwissenschaften; Nr. 36) 1892.

Ørsted, Hans Christian: Ansicht der chemischen Naturgesetze, durch die neueren Entdeckungen gewonnen. Berlin: Realschulbuchhandlung 1812. Ins Französische übersetzt von Serres de Mesplès, Marcel Pierre de Toussaint. Paris 1813. Reprint: Meyer, K. (Hg.): H. C. Ørsted Scientific Papers. Copenhagen 1920.

Pauli, Wolfgang: Über Gasentartung und Paramagnetismus. In: Zeitschrift für Physik 41 (1926), S. 81–102.

Peierls, Rudolf Ernst: Zur Theorie der galvanomagnetischen Effekte. In: Zeitschrift für Physik 53 (1929), S. 255–266.

Peierls, Rudolf Ernst: Zur kinetischen Theorie der Wärmeleitung in Kristallen. In: Annalen der Physik 395, (5) 3 (1929), S. 1055–1101.

Peltier, Jean: Nouvelles Expériences sur la Caloricité des courans électriques. In: Annales de Chimie et de Physique 56 (1834), S. 371–386.

Planck, Max: Gedächtnisrede auf Paul Drude (1863–1906), gehalten in der Sitzung der Deutschen Physikalischen Gesellschaft am 30. November 1906. In: Verhandlungen der Deutschen Physikalischen Gesellschaft 8 (1906), Nr. 23, 599–639.

Riecke, Eduard: Über die Bewegung eines elektrischen Teilchens in einem homogenen magnetischen Felde und das negative elektrische Glimmlicht. In: Annalen der Physik und Chemie, N. F., hrsg. von Gustav Wiedemann, (3) 13 (1881), S. 191–194.

Riecke, Eduard: Zur Theorie des Galvanismus und der Wärme. In: Annalen der Physik und Chemie, N. F., hrsg. von Gustav und Eilhard Wiedemann. (3) 66 (1898), S. 353–389 und 545–581, Nachtrag S. 1199–1200.

Riecke, Eduard: Über die Elektronentheorie des Galvanismus und der Wärme. In: Jahrbuch der Radioaktivität und Elektronik, Hrsg. von Joh. Stark, 3. Bd. 1906, Leipzig: S. Hirzel 1907, S. 24–47.

Riecke, Eduard: Die jetzigen Anschauungen über das Wesen des metallischen Zustandes. Vortrag auf der Hauptversammlung der Deutschen Bunsen-Gesellschaft, Mai 1909. In: Physikalische Zeitschrift 10 (1909), Nr. 15, S. 508–519.

Riecke, Eduard: Zur Elektronentheorie der thermoelektrischen und elektrothermischen Erscheinungen. In: Elster-Geitel Festschrift. Braunschweig: Friedrich Vieweg & Sohn 1915, S. 71–104.

Röseberg, Ulrich: Niels Bohr – Leben und Werk eines Atomphysikers. Berlin: Akademieverlag 1985.

Rosenberger, Ferdinand: Die Geschichte der Physik in Grundzügen mit synchronistischen Tabellen der Mathematik, der Chemie und beschreibenden Naturwissenschaften sowie der allgemeinen Geschichte. Bd. I (1882) Geschichte der Physik im Altertum und im Mittelalter, Bd. II (1884) Geschichte der Physik in der neueren Zeit, Bd. III Geschichte der Physik in den letzten hundert Jahren. Braunschweig: Vieweg (3 Teile in 2 Bänden) 1887–1890. (Reprint: Hildesheim: Georg Olms 1965).

Schmid, August: Eduard Riecke, Prof. der Experimentalphysik an der Universität Göttingen. In: Württembergischer Nekrolog für das Jahr 1915. Stuttgart 1919, S. 83–93.

Seebeck, Thomas Johann: Magnetische Polarisation der Metalle und Erze durch Temperaturdifferenz. Ergänzter Auszug aus vier Vorlesungen in der deutschen Akademie der Wissenschaften in Berlin 1821/22. Leipzig: Wilhelm Engelmann (Ostwalds Klassiker der exakten Wissenschaften; Band 70) 1895.

Seeliger, Rudolf: Elektronentheorie der Metalle. In: Encyklopädie der Mathematischen Wissenschaften, Bd. 5 (Physik), II. Teil (1921). Leipzig: B. G. Teubner Verlag 1904–1922.

Sommerfeld, Arnold: *Atombau und Spektrallinien.* Braunschweig: Vieweg 1919.

Sommerfeld, Arnold: Zur Elektronentheorie der Metalle auf Grund der Fermischen Statistik. In: Die Naturwissenschaften 15 (1927), Heft 4, S. 825–832.

Sommerfeld, Arnold: Zur Elektronentheorie der Metalle auf Grund der Fermi'schen Statistik. 1. Allgemeines, Strömungs- und Austrittsvorgänge. In: Zeitschrift für Physik 47 (1928) S. 1–32.

Sommerfeld, Arnold und Hans Bethe: Elektronentheorie der Metalle. In: Geiger/Scheel (Hg.): Handbuch der Physik, Volume 24, Part 2. Berlin: Springer 1933, S. 333–622.

Stark, Johannes: Prinzipien der Atomdynamik, 3 Bände. Leipzig 1910–1914.

Sutter, Paul: Die Elektronentheorie der Metalle mit besonderer Berücksichtigung der Theorie von Bohr und der galvano-magnetischen und thermomagnetischen Erscheinungen. Bern: P. Haupt 1920.

Teichmann, Jürgen: Zur Entwicklung von Grundbegriffen der Elektrizitätslehre insbesondere des elektrischen Stromes bis 1820. Hildesheim: Gerstenberg (Arbor scientiarum: Reihe A, Abhandlungen; Bd. 4) 1974.

Thomson, Joseph John: The Corpusculartheory of Metal. (London 1907) Deutsche Übersetzung von G. Siebert: Die Korpusculartheorie der Materie. Braunschweig: Friedrich Vieweg & Sohn 1908.

Thomson, William: On the Dynamical Theory of Heat: Part IV. Thermo-electric Currents. In: Transactions of the Royal Society of Edinburgh 21 (1854), S. 123–171.

Tolman, Richard C. and T. Dale Stewart: The acceleration of electrical conductors. In: Journal of the American Chemical Society 36 (1914), S. 466–485.

Tolman, Richard C. and T. Dale Stewart: The Electromotive Force Produced by the Acceleration of Metals. In: Physical Review (2) 8 (1916), No. 2, S. 97–116.

Tomaschek, Rudolf: Grimsehls Lehrbuch der Physik, zum Gebrauch beim Unterricht, neben akademischen Vorlesungen und zum Selbststudium. Band 1: Mechanik, Wärmelehre, Akustik. Leipzig, Berlin: B. G. Teubner (14., unveränderte Auflage) 1944.

Voigt, Woldemar: Eduard Riecke als Physiker. In: Physikalische Zeitschrift 16 (1915), No. 12, S. 219–221.

Weber, Wilhelm: Elektrodynamische Maassbestimmungen, insbesondere Widerstandsmessungen (1852). In: Weber Werke, Bd. 3, S. 301–471.

Weber, Wilhelm: Elektrodynamische Maassbestimmungen, insbesondere über das Princip der Erhaltung der Energie. In: Abh. d. Kgl. Sächs. Ges. d. Wiss., mathem.-phys. Kl. (1871) 10, Leipzig. In: Weber Werke, Bd. 4, S. 247–299.

Weber, Wilhelm: Zur Galvanometrie. In: Abh. d. Kgl. Gesellschaft der Wissenschaften zu Göttingen, mathem. Kl. (1862) 10, Göttingen. In: Weber Werke, Bd. 4, S. 17–96. Verkürzt schon vorher in Annalen der Physik und Chemie, hrsg. von J. C. Poggendorff, (2) 82 (1851), S. 337–369.

Weber Werke, Bd. 1–6. Hrsg. von der Kgl. Gesellschaft der Wissenschaften zu Göttingen. Berlin: Julius Springer 1892–1894.

Wiedemann, Gustav Heinrich und Rudolph Franz: Ueber die Wärme-Leitungsfähigkeit der Metalle. In: Annalen der Physik, hg. von Johann Christian Poggendorff, 165, (2) 89 (1853), Nr. 8, S. 497–531. Nachdruck: Wiedemann, Gustav Heinrich und Rudolph Franz: Über die Wärme-Leitungsfähigkeit der Metalle. Hg. von Arthur Wehnelt. Leipzig: Akademische Verlagsgesellschaft (Ostwald's Klassiker der Exakten Wissenschaften; 222) 1927.

Wiedemann, Gustav Heinrich: Die Lehre vom Galvanismus und Elektromagnetismus nebst ihren technischen Anwendungen. 2 Bände. Braunschweig: Vieweg 1861–1863.

Wiedemann, Gustav Heinrich: Die Lehre von der Elektricität. (5 Tle., 1883–1885). 4 Bände. Braunschweig: Vieweg (2. Auflage) 1895–1898. Dritter Band (1895), Elektrodynamik, S. 1–93, Vierter Band (1898), Hypothetische Ansichten über das Wesen und die Wirkungsweise der Elektricität, S. 797–1039.

Wiederkehr, Karl Heinrich: Wilhelm Webers Stellung in der Entwicklung der Elektrizitätslehre. Dissertation, Universität Hamburg 1960.

Wiederkehr, Karl Heinrich: Aus der Geschichte des Göttinger Magnetischen Vereins und seiner Resultate. In: Nachrichten der Akademie der Wissenschaften in Göttingen. II. Mathem.-Physikal. Klasse (1964), Nr. 14.

Wiederkehr, Karl Heinrich: Wilhelm Eduard Weber. Erforscher der Wellenbewegung und der Elektrizität 1804–1891. Stuttgart; Wissenschaftliche Verlagsgesellschaft (Große Naturforscher; Bd. 32, hrsg. von H. Degen) 1967.

Wiederkehr, Karl Heinrich: Über die Verleihung der Copley-Medaille an Gauß und die Mitarbeit Englands im Göttinger Magnetischen Verein. Mitteilungen der Gauß-Gesellschaft Nr. 19 (1982).

Wiederkehr, Karl Heinrich: Oersteds „Ansicht der chemischen Naturgesetze" 1812 und seine Naturphilosophischen Betrachtungen über Elektrizität und Magnetismus. In: Gesnerus 47 (1990), S. 161–183.

Wiederkehr, Karl Heinrich: Faradays Feldkonzept und Hans Christian Oersted. In: Physikalische Blätter 47 (1991), S. 825–830.

Wiederkehr, Karl Heinrich: Elektromagnetismus – Schlüsselphänomen für Faraday. In: Der Mathematische und Naturwissenschaftliche Unterricht (MNU) 44 (1991), S. 131–137.

Wiederkehr, Karl Heinrich: W. Weber und die Entwicklung der Elektrodynamik. In: II. Webersympossium. Die Gebrüder Weber – Wegbereiter der interdiziplinären Forschung. Halle und Leipzig Oktober und November 1993. Hrsg. vom Fachbereich Physik Martin-Luther-Universität Halle-Wittenberg und Wilhelm Weber Gesellschaft e. V., S. 39–54.

Wiederkehr, Karl Heinrich: Wilhelm Weber und Maxwells elektromagnetische Lichttheorie. In: Gesnerus 51 (1994), Part. 3/4, S. 256–267.

Wiederkehr, Karl Heinrich: Karl Friedrich Gauß (1777–1855) und Wilhelm Weber (1804–1891). In: Die großen Physiker. Hrsg. von Karl von Meÿenn. Bd. 1, München 1997, S. 357–370.

Wiederkehr, Karl Heinrich: Die Entdeckung des Elektrons. In: Der Mathematische und Naturwissenschaftliche Unterricht (MNU) 52 (1999), S. 131–139.

Wiederkehr, Karl Heinrich: Ein bisher unveröffentlicher Brief von Rudolf Kohlrausch an Andreas v. Ettingshausen (1796–1878) von 1854, das Kohlrausch-Weber-Experiment von 1854/55 und die Lichtgeschwindigkeit in Wilhelm Webers Elektrodynamik. In: NTM 12 (2004), S. 129–145.

Wiederkehr, Karl Heinrich: Zum Lebenswerk von Wilhelm Weber (1804- 1891). In: Wolfschmidt, Gudrun (Hg.): Vom Magnetismus zur Elektrodynamik. Herausgegeben anläßlich des 200. Geburtstages von Wilhelm Weber (1804–1891) und des 150. Todestages von Carl Friedrich Gauß (1777–1855). Katalog zur Ausstellung in der Staatsbibliothek Hamburg, 3. März bis 2. April 2005. Hamburg 2005, S. 73–91.

Wien, Wilhelm Zur Theorie der elektrischen Leitung in Metallen. In: Sitzungsberichte der königlich preußischen Akademie der Wissenschaften (1913), 1. Halbband. Berlin: Verlag der Akademie 1913, S. 184–200.

Wien, Wilhelm und F. Harms (Hg.): Handbuch der Experimentalphysik. 26 Bände. Leipzig: Akademische Verlagsgesellschaft 1926–1937.

Figure 3.1:
Wilhelm Eduard Weber (1804–1891)
Lenard 1930, p. 230.

Works by Assis related to Weber's Law Applied to Electromagnetism and Gravitation

Andre Koch Torres Assis (Campinas, SP, Brazil)

We list the books and papers written by Andre Koch Torres Assis related with Weber's law applied to electromagnetism and gravitation.

3.1 Books

Books by Andre Koch Torres Assis related with Weber's law applied to electromagnetism and gravitation:

- Books in English: [1, 2, 3, 4].
- Books in Portuguese: [5, 6, 7, 8, 9, 10].

3.2 Papers

All papers by Andre Koch Torres Assis are available in PDF format at:

http://www.ifi.unicamp.br/~assis.

Specific papers related with Weber's law applied to electromagnetism and gravitation:

- Papers in English: [11, 12, 13, 14, 15, 16, 17, 18, 19, 20, 21, 22, 23, 24, 25, 26, 27, 28, 29, 30, 31, 32, 33, 34, 35, 36, 37, 38, 39, 40, 41, 42, 43, 44, 45, 46, 47, 48, 49, 50, 51, 52, 53, 54, 55, 56, 57, 58, 59, 60, 61, 62, 63, 64, 65, 66, 67, 68, 69, 70, 71, 72, 73, 74, 75, 76, 77, 78, 79, 80, 81, 82, 83, 84, 85, 86, 87, 88].

- Papers in Portuguese: [89, 90, 91, 92, 93, 94, 95, 96, 97, 98, 99, 100, 101, 102, 103, 104, 105].

Bibliography

[1] ASSIS, ANDRE KOCH TORRES: *Weber's Electrodynamics*. Dordrecht: Kluwer Academic Publishers 1994. ISBN: 0-7923-3137-0.

[2] ASSIS, ANDRE KOCH TORRES: *Relational Mechanics*. Montreal: Apeiron 1999. ISBN: 0-9683689-2-1. Available at: http://www.ifi.unicamp.br/~assis.

[3] BUENO, M. D. A. AND ANDRE KOCH TORRES ASSIS: *Inductance and Force Calculations in Electrical Circuits*. Huntington, New York: Nova Science Publishers 2001. ISBN: 1-56072-917-1.

[4] ASSIS, ANDRE KOCH TORRES AND J. A. HERNANDES: *The Electric Force of a Current: Weber and the Surface Charges of Resistive Conductors Carrying Steady Currents*. Montreal: Apeiron 2007.
ISBN: 978-0-9732911-5-5. Available at: http://www.ifi.unicamp.br/~assis.

[5] ASSIS, ANDRE KOCH TORRES: *Curso de Eletrodinâmica de Weber*. Campinas: Setor de Publicações do Instituto de Física da Universidade Estadual de Campinas – Unicamp 1992.
Notas de Física IFGW Número 5. Available at: http://www.ifi.unicamp.br/~assis.

[6] ASSIS, ANDRE KOCH TORRES: *Eletrodinâmica de Weber – Teoria, Aplicações e Exercícios*. Campinas: Editora da Universidade Estadual de Campinas – UNICAMP 1995. ISBN: 85-268-0358-1.

[7] ASSIS, ANDRE KOCH TORRES: *Mecânica Relacional*. Campinas: Editora do Centro de Lógica, Epistemologia e História da Ciência da UNICAMP/FAPESP 1998.
ISBN: 85-86497-01-0. Available at: http://www.ifi.unicamp.br/~assis.

[8] MARCELO BUENO AND ANDRE KOCH TORRES ASSIS: *Cálculo de Indutância e de Força em Circuitos Elétricos*. Florianópolis, Maringá: Editora da UFSC/Editora da UEM 1998. ISBN: 85-328-0119-6.

[9] ASSIS, ANDRE KOCH TORRES: *Uma Nova Física*. São Paulo: Editora Perspectiva 1999. ISBN: 85-273-0199-7.

[10] ASSIS, ANDRE KOCH TORRES AND J. A. HERNANDES: *A Força Elétrica de uma Corrente: Weber e as Cargas Superficiais de Condutores Resistivos com Correntes Constantes*. São Paulo and Maceió: Edusp and Edufal (*Acadêmica*; vol. 73) 2009. ISBNs: 978-85-314-1123-6 and 978-85-7177-431-5.

[11] Assis, Andre Koch Torres: On Mach's principle. In: *Foundations of Physics Letters* 2 (1989), p. 301–318.

[12] Assis, Andre Koch Torres: Weber's law and mass variation. In: *Physics Letters A* 136 (1989), p. 277–280.

[13] Assis, Andre Koch Torres: Modern experiments related to Weber's electrodynamics. In: Bartocci, U. and J. P. Wesley (ed.): *Proceedings of the Conference on Foundations of Mathematics and Physics*. Blumberg, Germany: Benjamin Wesley Publisher 1990, p. 8–22.

[14] Assis, Andre Koch Torres: Deriving Ampère's law from Weber's law. In: *Hadronic Journal* 13 (1990), p. 441–451.

[15] Assis, Andre Koch Torres: Can a steady current generate an electric field? In: *Physics Essays* 4 (1991), p. 109–114.

[16] Clemente, R. A. and Andre Koch Torres Assis: Two-body problem for Weber-like interactions. In: *International Journal of Theoretical Physics* 30 (1991), p. 537–545.

[17] Assis, Andre Koch Torres and J. J. Caluzi: A limitation of Weber's law. In: *Physics Letters A* 160 (1991), p. 25–30.

[18] Assis, Andre Koch Torres and R. A. Clemente: The ultimate speed implied by theories of Weber's type. In: *International Journal of Theoretical Physics* 31 (1992), p. 1063–1073.

[19] Assis, Andre Koch Torres: On the mechanism of railguns. In: *Galilean Electrodynamics* 3 (1992), p. 93–95.

[20] Assis, Andre Koch Torres: On forces that depend on the acceleration of the test body. In: *Physics Essays* 5 (1992), p. 328–330.

[21] Assis, Andre Koch Torres: Centrifugal electrical force. In: *Communications in Theoretical Physics* 18 (1992), p. 475–478.

[22] Assis, Andre Koch Torres: Compliance of a Weber's force law for gravitation with Mach's principle. In: Kropotkin, P. N. et al. (ed.): *Space and Time Problems in Modern Natural Science, Part II*. St.-Petersburg: Tomsk Scientific Center of the Russian Academy of Sciences (Series: "The Universe Investigation Problems"; Issue 16) 1993, p. 263–270.

[23] Assis, Andre Koch Torres: Changing the inertial mass of a charged particle. In: *Journal of the Physical Society of Japan* 62 (1993), p. 1418–1422.

[24] Assis, Andre Koch Torres and R. A. Clemente: The influence of temperature on gravitation. In: *Il Nuovo Cimento B* 108 (1993), p. 713–716.

[25] Graneau, P. and Andre Koch Torres Assis: Kirchhoff on the motion of electricity in conductors. In: *Apeiron* 19 (1994), p. 19–25.

[26] Assis, Andre Koch Torres and D. S. Thober: Unipolar induction and Weber's electrodynamics. In: Barone, M. and F. Selleri (ed.): *Frontiers of Fundamental Physics*. New York: Plenum Press 1994, p. 409–414.

[27] Assis, Andre Koch Torres: Acceleration dependent forces: reply to Smulsky. In: *Apeiron* 2 (1995), p. 25.

[28] Assis, Andre Koch Torres and P. Graneau: The reality of Newtonian forces of inertia. In: *Hadronic Journal* 18 (1995), p. 271–289.

[29] Caluzi, J. J. and Andre Koch Torres Assis: Schrödinger's potential energy and Weber's electrodynamics. In: *General Relativity and Gravitation* 27 (1995), p. 429–437.

[30] Bueno, Marcelo A. and Andre Koch Torres Assis: A new method for inductance calculations. In: *Journal of Physics D* 28 (1995), p. 1802–1806.

[31] Assis, Andre Koch Torres: Weber's law and Mach's principle. In: Barbour, J. B. and H. Pfister (ed.): *Mach's Principle – From Newton's Bucket to Quantum Gravity*. Boston: Birkhäuser 1995, p. 159–171.

[32] Assis, Andre Koch Torres: Weber's force versus Lorentz's force. In: *Physics Essays* 8 (1995), p. 335–341.

[33] Assis, Andre Koch Torres: Gravitation as a fourth order electromagnetic effect. In: Barrett, T. W. and D. M. Grimes (ed.): *Advanced Electromagnetism: Foundations, Theory and Applications*. Singapore: World Scientific 1995, p. 314–331.

[34] Assis, Andre Koch Torres and Marcelo Bueno: Longitudinal forces in Weber's electrodynamics. In: *International Journal of Modern Physics B* 9 (1995), p. 3689–3696.

[35] Assis, Andre Koch Torres and Marcelo A. Bueno: Equivalence between Ampère and Grassmann's forces. In: *IEEE Transactions on Magnetics* 32 (1996), p. 431–436.

[36] Assis, Andre Koch Torres and P. Graneau: Nonlocal forces of inertia in cosmology. In: *Foundations of Physics* 26 (1996), p. 271–283.

[37] Caluzi, J. J. and Andre Koch Torres Assis: An analysis of Phipps's potential energy. In: *Journal of the Franklin Institute B* 332 (1995), p. 747–753.

[38] Caluzi, J. J. and Andre Koch Torres Assis: The oscillatory motion of charged particles by Weber's electrodynamics. In: Gill, T. P. (ed.): *New Frontiers in Physics, Volume I*. Palm Harbor: Hadronic Press 1996, p. 129–143.

[39] Assis, Andre Koch Torres: Circuit theory in Weber electrodynamics. In: *European Journal of Physics* 18 (1997), p. 241–246.

[40] Bueno, Marcelo and Andre Koch Torres Assis: Equivalence between the formulas for inductance calculation. In: *Canadian Journal of Physics* 75 (1997), p. 357–362.

[41] BUENO, MARCELO AND ANDRE KOCH TORRES ASSIS: Proof of the identity between Ampère and Grassmann's forces. In: *Physica Scripta* 56 (1997), p. 554–559.

[42] BUENO, MARCELO AND ANDRE KOCH TORRES ASSIS: Self-inductance of solenoids, bi-dimensional rings and coaxial cables. In: *Helvetica Physica Acta* 70 (1997), p. 813–821.

[43] CALUZI, J. J. AND ANDRE KOCH TORRES ASSIS: A critical analysis of Helmholtz's argument against Weber's electrodynamics. In: *Foundations of Physics* 27 (1997), p. 1445–1452.

[44] ALMEIDA BUENO, MARCELO DE AND ANDRE KOCH TORRES ASSIS: Deriving force from inductance. In: *IEEE Transactions on Magnetics* 34 (1998), p. 317–319.

[45] ASSIS, ANDRE KOCH TORRES AND J. I. CISNEROS: The problem of surface charges and fields in coaxial cables and its importance for relativistic physics. In: SELLERI, F. (ed.): *Open Questions in Relativistic Physics*. Montreal: Apeiron 1998, p. 177–185.

[46] ASSIS, ANDRE KOCH TORRES; RODRIGUES JR., W. A. AND A. J. MANIA: The electric field outside a stationary resistive wire carrying a constant current. In: *Foundations of Physics* 29 (1999), p. 729–753.

[47] ASSIS, ANDRE KOCH TORRES AND J. J. CALUZI: Charged particle oscillating near a capacitor. In: *Galilean Electrodynamics* 10 (1999), p. 103–106.

[48] ASSIS, ANDRE KOCH TORRES AND A. J. MANIA: Surface charges and electric field in a two-wire resistive transmission line. In: *Revista Brasileira de Ensino de Física* 21 (1999), p. 469–475.

[49] ASSIS, ANDRE KOCH TORRES: Arguments in favour of action at a distance. In: CHUBYKALO, A. E.; POPE, V. AND R. SMIRNOV-RUEDA (ed.): *Instantaneous Action at a Distance in Modern Physics – "Pro" and "Contra"*. Commack: Nova Science Publishers 1999, p. 45–56.

[50] ASSIS, ANDRE KOCH TORRES: The meaning of the constant c in Weber's electrodynamics. In: MONTI, R. (ed.): *Proc. of the Int. Conf. Galileo Back in Italy II*. Bologna: Soc. Ed. Andromeda 2000, p. 23–36.

[51] ASSIS, ANDRE KOCH TORRES AND J. I. CISNEROS: Surface charges and fields in a resistive coaxial cable carrying a constant current. In: *IEEE Transactions on Circuits and Systems I* 47 (2000), p. 63–66.

[52] ASSIS, ANDRE KOCH TORRES, J. FUKAI, AND H. B. CARVALHO: Weberian induction. In: *Physics Letters A* 268 (2000), p. 274–278.

[53] ASSIS, ANDRE KOCH TORRES AND J. GUALA-VALVERDE: Mass in relational mechanics. In: *Apeiron* 7 (2000), p. 131–132.

[54] ASSIS, ANDRE KOCH TORRES AND H. TORRES SILVA: Comparison between Weber's electrodynamics and classical electrodynamics. In: *Pramana Journal of Physics* 55 (2000), p. 393–404.

[55] ASSIS, ANDRE KOCH TORRES AND MARCELO BUENO: Bootstrap effect in classical electrodynamics. In: *Revista Facultad de Ingenieria de la Universidad de Tarapaca (Chile)* 7 (2000), p. 49–55.

[56] ASSIS, ANDRE KOCH TORRES: Comment on "Experimental proof of standard electrodynamics by measuring the self-force on a part of a current loop". In: *Physical Review E* 62 (2000), p. 7544.

[57] ASSIS, ANDRE KOCH TORRES: On the propagation of electromagnetic signals in wires and coaxial cables according to Weber's electrodynamics. In: *Foundations of Physics* 30 (2000), p. 1107–1121.

[58] ASSIS, ANDRE KOCH TORRES AND A. ZYLBERSZTAJN: The influence of Ernst Mach in the teaching of mechanics. In: *Science and Education* 10 (2001), p. 137–144.

[59] ASSIS, ANDRE KOCH TORRES: Applications of the principle of physical proportions to gravitation. In: RUDNICKI, K. (ed.): *Gravitation, Electromagnetism and Cosmology – Toward a New Synthesis*. Montreal: Apeiron 2001, p. 1–7.

[60] ASSIS, ANDRE KOCH TORRES; HERNANDES, J. A. AND J. E. LAMESA: Surface charges in conductor plates carrying constant currents. In: *Foundations of Physics* 31 (2001), p. 1501–1511.

[61] TORRES S., H. AND ANDRE KOCH TORRES ASSIS: The influence of the electric field outside a resistive solenoid on the Aharonov-Bohm effect. In: *Revista de la Facultad de Ingenieria de la Universidad de Tarapaca (Chile)* 9 (2001), p. 29–34.

[62] ASSIS, ANDRE KOCH TORRES: On the unification of forces of nature. In: *Annales de la Fondation Louis de Broglie*, 27 (2002), p. 149–161.

[63] ASSIS, ANDRE KOCH TORRES, K. REICH, AND K. H. WIEDERKEHR: Gauss and Weber's creation of the absolute system of units in physics. In: *21st Century* 15 (2002), No. 3, p. 40–48.

[64] HERNANDES, J. A. AND ANDRE KOCH TORRES ASSIS: Potential, electric field and surface charges for a resistive long straight strip carrying a constant current. In: SBF (ed.): *Proceedings of the XXIII Encontro Nacional de Física de Partículas e Campos*. São Paulo: Sociedade Brasileira de Física 2002p. P–163. Published at (accessed in 2008): http://www.sbf1.if.usp.br/eventos/enfpc/xxiii/procs/RES80.pdf.

[65] ASSIS, ANDRE KOCH TORRES AND J. A. HERNANDES: Electric potential for a toroidal ring carrying a constant current. In SBF (ed.): *Proceedings of the XXIII Encontro Nacional de Física de Partículas e Campos*. São Paulo: Sociedade Brasileira de Física 2002, p. P–115. Published at (accessed in 2008): http://www.sbf1.if.usp.br/eventos/enfpc/xxiii/procs/RES81.pdf.

[66] ASSIS, ANDRE KOCH TORRES AND J. GUALA-VALVERDE: On absolute and relative motions in physics. In: *Journal of New Energy* 6 (2002), p. 8–12.

[67] ASSIS, ANDRE KOCH TORRES AND J. GUALA-VALVERDE: Frequency in relational mechanics. In: *Annales de la Fondation Louis de Broglie.* 28 (2003), p. 83–97.

[68] ASSIS, ANDRE KOCH TORRES: The relationship between Mach's principle and the principle of physical proportions. In: SACHS, M. AND A. R. ROY (ed.): *Mach's Principle and the Origin of Inertia.* Montreal: Apeiron 2003, p. 37–44.

[69] HERNANDES, J. A. AND ANDRE KOCH TORRES ASSIS: The potential, electric field and surface charges for a resistive long straight strip carrying a steady current. In: *American Journal of Physics* 71 (2003), p. 938–942.

[70] ASSIS, ANDRE KOCH TORRES AND K. H. WIEDERKEHR: Weber quoting Maxwell. In: *Mitteilungen der Gauss-Gesellschaft* 40 (2003), p. 53–74.

[71] J. FUKAI AND ANDRE KOCH TORRES ASSIS: Testing Mach's principle in electrodynamics. In: *Canadian Journal of Physics* 81 (2003), p. 1239–1242.

[72] HERNANDES, J. A. AND ANDRE KOCH TORRES ASSIS: Electric potential for a resistive toroidal conductor carrying a steady azimuthal current. In: *Physical Review E* 68 (2003), 046611.

[73] ASSIS, ANDRE KOCH TORRES: On the first electromagnetic measurement of the velocity of light by Wilhelm Weber and Rudolf Kohlrausch. In: BEVILACQUA, F. AND E. A. GIANNETTO (ed.): *Volta and the History of Electricity.* Milano: Università degli Studi di Pavia and Editore Ulrico Hoepli 2003, p. 267–286.

[74] ASSIS, ANDRE KOCH TORRES: The principle of physical proportions. In: *Annales de la Fondation Louis de Broglie*, 29 (2004), p. 149–171.

[75] ASSIS, ANDRE KOCH TORRES; REICH, K. AND K. H. WIEDERKEHR: On the electromagnetic and electrostatic units of current and the meaning of the absolute system of units – For the 200th anniversary of Wilhelm Weber's birth. In: *Sudhoffs Archiv* 88 (2004), p. 10–31.

[76] HERNANDES, J. A. AND ANDRE KOCH TORRES ASSIS: Surface charges and external electric field in a toroid carrying a steady current. In: *Brazilian Journal of Physics* 34 (2004), p. 1738–1744.

[77] HERNANDES, J. A.; CAPELAS DE OLIVEIRA, E. AND ANDRE KOCH TORRES ASSIS: Potential, electric field and surface charges close to the battery for a resistive cylindrical shell carrying a steady longitudinal current. In: *Revista de la Facultad de Ingeniería (Chile)* 12 (2004), p. 13–20.

[78] MENDES, R. S.; MALACARNE, L. C. AND ANDRE KOCH TORRES ASSIS: Virial theorem for Weber's law. In: CHUBYKALO, A.; ESPINOZA, A.; ONOOCHIN, V. AND R. SMIRNOV-RUEDA (ed.): *Has the Last Word Been Said on Classical Electrodynamics? New Horizons.* Paramus: Rinton Press 2004, p. 67–70.

[79] ASSIS, ANDRE KOCH TORRES: Weber's electrodynamics and Mach's principle in the 21st century. In: WOLFSCHMIDT, GUDRUN (ed.): *Vom Magnetismus zur Elektrodynamik*. Hamburg: Schwerpunkt Geschichte der Naturwissenschaften, Mathematik und Technik 2005, p. 10–11.

[80] ASSIS, ANDRE KOCH TORRES AND J. A. HERNANDES: Telegraphy equation from Weber's electrodynamics. In: *IEEE Transactions on Systems and Circuits II* 52 (2005), p. 289–292.

[81] HERNANDES, J. A. AND ANDRE KOCH TORRES ASSIS: The electric field outside and inside a resistive spherical shell carrying a steady azimuthal current. In: *Physica Scripta* 72 (2005), p. 212–217.

[82] HERNANDES, J. A.; CAPELAS DE OLIVEIRA, E. AND ANDRE KOCH TORRES ASSIS: Resistive plates carrying a steady current: electric potential and surface charges close to the battery. In: *Foundations of Physics Letters* 18 (2005), p. 275–289.

[83] HERNANDES, J. A. AND ANDRE KOCH TORRES ASSIS: Electric potential due to an infinite conducting cylinder with internal or external point charge. In: *Journal of Electrostatics* 63 (2005), p. 1115–1131.

[84] ASSIS, ANDRE KOCH TORRES AND J. A. HERNANDES: Magnetic energy and effective inertial mass of the conduction electrons in circuit theory. In: *Electromagnetic Phenomena* 6 (2006), p. 31–35.

[85] WEBER, W.: Determinations of electrodynamic measure: concerning a universal law of electrical action. In: *21st Century Science & Technology*, posted March 2007, translated by S. P. JOHNSON, edited by L. HECHT AND A. K. T. ASSIS.
Original paper: Elektrodynamische Maassbestimmungen: Ueber ein allgemeines Grundgesetz der elektrischen Wirkung, Treatise at the founding of the Royal Scientific Society of Saxony on the day of the 200th anniversary celebration of Leibniz's birthday. Leipzig: Prince Jablonowski Society 1846, p. 211–378; reprinted in *Wilhelm Weber's Werke, Vol. III: Galvanismus und Elektrodynamik, part 1*, edited by H. WEBER. Berlin: Julius Springer Verlag 1893, p. 25–214. Available at: http://www.21stcenturysciencetech.com/.

[86] HERNANDES, J. A.; MANIA, A. J.; LUNA, F. R. T. AND ANDRE KOCH TORRES ASSIS: The internal and external electric fields for a resistive toroidal conductor carrying a steady poloidal current. In: *Physica Scripta* 78 (2008), p. 015403.

[87] HERNANDES, J. A. AND ANDRE KOCH TORRES ASSIS: Surface charges and fields in stationary conductors with steady currents. In: SIDHARTH, B. G.; HONSELL, F.; MANSUTTI, O.; SREENIVASAN, K. AND A. DE ANGELIS (ed.): *Frontiers of Fundamental and Computational Physics*. Proceedings of the 9th International Symposium (Udine and Trieste, Italy, 7-9 January 2008). New York: Melville (AIP Conference Proceedings; Vol. 1018) 2008, p. 236–239.

[88] WEBER, W.: Determinations of electrodynamic measure: particularly in respect to the connection of the fundamental laws of electricity with the law of gravitation. In: *21st Century Science & Technology*, posted November 2008, translated by G. GREGORY, edited by L. HECHT AND A. K. T. ASSIS.
Original paper: Elektrodynamische Maassbestimmungen insbesondere über den Zusammenhang des elektrischen Grundgesetzes mit dem Gravitationsgesetze. This work was originally published in Wilhelm Weber's Werke, Vol. IV: Galvanismus und Elektrodynamik, part 2, edited by H. WEBER. Berlin: Julius Springer Verlag 1894, p. 479–525. Available at: http://www.21stcenturysciencetech.com/.

[89] ASSIS, ANDRE KOCH TORRES: Wilhelm Eduard Weber (1804-1891) – Sua vida e sua obra. In: *Revista da Sociedade Brasileira de História da Ciência* 5 (1991), p. 53–59.

[90] ASSIS, ANDRE KOCH TORRES: A eletrodinâmica de Weber e seus desenvolvimentos recentes. In: *Ciência e Natura* 17 (1995), p. 7–16.

[91] XAVIER JR., A. L. AND ANDRE KOCH TORRES ASSIS: O cumprimento do postulado de relatividade na mecânica clássica – uma tradução comentada de um texto de Erwin Schrödinger sobre o princípio de Mach. In: *Revista da Sociedade Brasileira de História da Ciência* 12 (1994), p. 3–18.

[92] XAVIER JR., A. L. AND ANDRE KOCH TORRES ASSIS: Schrödinger, Reissner, Weber e o princípio de Mach. In: *Revista da Sociedade Brasileira de História da Ciência* 17 (1997), p. 103–106.

[93] ASSIS, ANDRE KOCH TORRES: Comparação entre as eletrodinâmicas de Weber e de Maxwell-Lorentz. In: *Episteme* 3 (1998), p. 7–15.

[94] ASSIS, ANDRE KOCH TORRES: O conceito de massa na mecânica relacional e na relatividade geral. In: FERNANDES, H. C. C. (ed.): *Anais do IV Simpósio de Pesquisa e Extensão em Tecnologia*. Natal: Centro de Tecnologia – UFRN 1999, p. 191–192.

[95] A. ZYLBERSZTAJN AND ANDRE KOCH TORRES ASSIS: Sobre a possível realidade das forças fictícias: uma visão relacional da mecânica. In: *Acta Scientiarum* 21 (1999), p. 817–822.

[96] ASSIS, ANDRE KOCH TORRES AND J. A. HERNANDES: A repulsão coulombiana não explica a explosão de fios. In: *Acta Scientiarum* 21 (1999), p. 837–839.

[97] ASSIS, ANDRE KOCH TORRES AND J. A. HERNANDES: Cargas superficiais em placas condutoras com correntes constantes. In: SBF (ed.): *Anais do XX Encontro Nacional de Física de Partículas e Campos, São Lourenço, 25 a 29/10/99*. São Paulo: Sociedade Brasileira de Física 1999, page 1. Published at (accessed in 2008): http://www.sbf1.if.usp.br/eventos/enfpc/xx/procs/res224/.

[98] ASSIS, ANDRE KOCH TORRES: A primeira medida eletromagnética da velocidade da luz por Weber e Kohlrausch. In: GOLDFARB, J. L. AND M. H. M. FERRAZ (ed.): *Anais do VII Seminário Nacional de História da Ciência e da Tecnologia*. São Paulo: Editora da USP/Editora da Unesp/Imprensa Oficial do Estado/Soc. Bras. de Hist. da Ciência 2000, p. 65–71.

[99] HERNANDES, J. A. AND ANDRE KOCH TORRES ASSIS: Propagação de sinais em condutores com a eletrodinâmica de Weber e comparação com o eletromagnetismo clássico. In: SBF (ed.): *Anais do XXI Encontro Nacional de Física de Partículas e Campos, São Lourenço, 23 a 27/10/00*. São Paulo: Sociedade Brasileira de Física 2000, page 1. Published at (accessed in 2008): http://www.sbf1.if.usp.br/eventos/enfpc/xxi/procs/res89/.

[100] ASSIS, ANDRE KOCH TORRES: Comparação entre a mecânica relacional e a relatividade general de Einstein. In: PESSOA JR., O. (ed.): *Fundamentos da Física 2 – Simpósio David Bohm*. São Paulo: Editora Livraria da Física 2001, p. 27–38.

[101] HERNANDES, J. A. AND ANDRE KOCH TORRES ASSIS: Propagação de sinais em condutores segundo a eletrodinâmica de Weber. In: *Ciência e Natura* 23 (2001), p. 7–26.

[102] ASSIS, ANDRE KOCH TORRES AND OSVALDO PESSOA JR.: Erwin Schrödinger e o princípio de Mach. In: *Cadernos de História e Filosofia da Ciência* 11 (2001), n. 2, p. 131–152.

[103] ASSIS, ANDRE KOCH TORRES: Tradução de uma obra de Gauss. In: *Revista Brasileira de Ensino de Física* 25 (2003), p. 226–249. Paper in Portuguese: "Translation of a work by Gauss": C. F. GAUSS: Die Intensität der erdmagnetischen Kraft, zurückgeführt auf absolutes Maass. In: *Annalen der Physik und Chemie* 28 (= 104), number 6, p. 241–273 and number 8, p. 591–615 (1833); and C. F. GAUSS: Die Intensität der erdmagnetischen Kraft auf absolutes Maass zurückgeführt. Ed. by DORN, E.. Leipzig: Wilhelm Engelmann Verlag (Ostwald's Klassiker der exakten Wissenschaften; Vol. 53) 1894. Translation by Kiel, notes by E. DORN. Portuguese translation by A. K. T. Assis.

[104] ASSIS, ANDRE KOCH TORRES: Interações na física — ação à distância versus ação por contato. In: SILVA, C. C. (ed.): *Estudos de História e Filosofia das Ciências: Subsídios para Aplicação no Ensino*. São Paulo: Editora Livraria da Física 2006, p. 87–102.

[105] WEBER, W. AND R. KOHLRAUSCH: Sobre a quantidade de eletricidade que flui através da seção reta do circuito em correntes galvânicas. In: *Revista Brasileira de História da Ciência* 1 (2008), p. 94–102. Portuguese translation by A. K. T. ASSIS of a paper by W. WEBER AND R. KOHLRAUSCH: "On the amount of electricity which flows through the cross-section of the circuit in galvanic currents." In: *Annalen der Physik und Chemie* 99 (1856), p. 10–25.

"Über die Elektricitätsmenge, welche bei galvanischen Strömen durch den Querschnitt der Kette fliesst." Reprinted in *Wilhelm Weber's Werke*, Vol. 3, ed. by H. WEBER. Berlin: Springer 1893, p. 597–608.

Figure 3.2:
Johann Karl Friedrich Zöllner (1834–1882)
http://www.geophys.tu-bs.de/geschichte/zoellner.html

Figure 3.3:
Annalen der Physik (4. Serie), Band 306 (1900), ed. by Paul Drude

4.1 Zeittafel zu Wilhelm Weber (1795–1878)

1804	24. Oktober, Wilhelm Eduard Weber, geboren zu Wittenberg. Vater, Michael Weber, Universitätsprediger in Halle (Saale), zuletzt Professor der Theologie. Mutter, Christiane Friederike Wilhelmine, geb. Lippold, † 1816, danach von der zweiten Ehefrau Friederike Henriette, geb. Pallas erzogen.
1825	Erstlingsarbeit in Gemeinschaft mit seinem älteren Bruder Ernst Heinrich Weber (1795–1878), Professor der Physiologie, Leipzig: Wellenlehre auf Experimente gegründet.
1826	Dissertation: *Theoria efficaciae laminarum maxime mobilium arcteque tubos aërem sonantem continentes claudentium.* (Theorie der Wirksamkeit höchst beweglicher Zungen, welche Röhren, die tönende Luft enthalten, eng verschließen).
1827	Habilitationsschrift: *Leges oscillationis oriundae, si duo corpora diversa celeritate oscillantia ita conjunguntur ut oscillare non possint nisi simul synchronice* (Gesetze der Schwingung, die entsteht, wenn zwei Körper von verschiedener Schwingungszahl so gekoppelt werden, daß sie nur gleichseitig und im Gleichtakt miteinander schwingen können).
1828	Außerordentlicher Professor an der Universität Halle. Erste Begegnung mit Carl Friedrich Gauß auf der Versammlung Deutscher Naturforscher und Ärzte zu Berlin.
1831	Berufung auf den freigewordenen Lehrstuhl der Physik in Göttingen (Gutachten von Gauß).
1837	Göttinger Sieben. Amtsenthebung
1843	Berufung auf den Physiklehrstuhl in Leipzig
1846	*Elektrodynamische Maßbestimmungen. Über ein Grundgesetz der elektrischen Wirkung* (Webersches Gesetz)
1849	Rückkehr nach Göttingen. Schaffung der Grundlagen für die absoluten Maßsysteme (absolutes elektromagnetisches, absolutes elektrodynamisches, absolutes elektrostatisches und des damit in Verbindung stehenden mechanischen Maßsystems); ferner Konstruktion geeigneter Meßgeräte
1855	Zusammen mit Rudolf Kohlrausch Bestimmung des Verhältnisses von absoluter elektrostatischer und absoluter elektromagnetischer Ladung. Maxwell erkennt 1861 darin die Lichtgeschwindigkeit, Grundlage für seine elektromagnetische Lichttheorie.
1881	Erster internationaler Elektrikerkongreß in Paris. Das absolute elektromagnetische Maßsystem Webers dient als Grundlage für die internationalen praktischen Maßeinheiten wie Volta, Ampère und Ohm.
1891	23. Juni gestorben. Auf dem Stadtfriedhof in Göttingen begraben.

Arbeiten zu Gauss und Weber
von K. H. Wiederkehr

Karl Heinrich Wiederkehr (Hamburg)

WIEDERKEHR, KARL HEINRICH: Hamburgs patriotische Bürger und die Göttinger Sieben. Vom Kampf der Hamburgischen Presse gegen die Zensur. In: *Hamburgische Geschichts- und Heimatblätter* 21 (1964), S. 197–208.

WIEDERKEHR, KARL HEINRICH: Wilhelm Webers Quasi-Elektronentheorie der vormaxwellschen Epoche. In: *Abhandlungen und Verhandlungen des Naturwissenschaftlichen Vereins in Hamburg*, N.F. Bd. VII (1962), Hamburg 1963.

WIEDERKEHR, KARL HEINRICH: Aus der Geschichte des Göttinger Magnetischen Vereins und seine Resultate. In: *Nachrichten der Akademie der Wissenschaften in Göttingen*, II. Mathematisch-Physikalische Klasse (1964), Nr. 14, Göttingen 1964, S. 165–205.

WIEDERKEHR, KARL HEINRICH: Wilhelm Weber, Kurzbiografie. In: Große Naturwissenschaftler. Hrsg. von ADOLF MEYER-ABICH UND FRITZ KRAFFT. Fischerbücherei 1970, neu bearbeitete und erweiterte Auflage, hrsg. von F. KRAFFT. Düsseldorf: VDI 1986.

WIEDERKEHR, KARL HEINRICH: Das bisher unbekannte Gauß-Gutachten zur Wiederbesetzung des Göttinger Physiklehrstuhls 1831. In: *Mitteilungen der Gauss-Gesellschaft Göttingen*, Nr. 10 (1973), S. 32–47.

WIEDERKEHR, KARL HEINRICH: Wilhelm Weber und die Entwicklung in der Geomagnetik und Elektrodynamik. In: *I. Weber-Symposium anlässlich des 100. Todestages von W. Weber in Halle und Wittenberg am 20. und 21. Juni 1991*. Martin-Luther-Universität Halle-Wittenberg, Fachbereich Physik und Wilhelm-Weber-Gesellschaft e.V., S. 1–14.

WIEDERKEHR, KARL HEINRICH: W. Weber und die Entwicklung der Elektrodynamik. In: *II. Weber-Symposium. Die Gebrüder Weber – Wegbereiter interdisziplinärer Forschung, in Halle und Leipzig am 16. Oktober und 18. November 1993*. Fachbereich Physik der Martin-Luther-Universität Halle-Wittenberg und

Wilhelm-Weber-Gesellschaft S. 39–54.

Auch in Mitteilungen der Gauss-Gesellschaft Göttingen, Nr. 29 (1992), S. 63–72. Mit einem Bild der drei Gebrüder Weber.

Dazu auch das Heft: Die Gebrüder Weber, zur Wanderausstellung der Weber-Gesellschaft.

WIEDERKEHR, KARL HEINRICH: Feier zum 190. Geburtstag von Wilhelm Weber und Aktivitäten der Weber-Gesellschaft. In: *Mitteilungen der Gauss-Gesellschaft Göttingen*, Nr. 32 (1995), S. 77–78.

WIEDERKEHR, KARL HEINRICH: Carl Friedrich Gauß (1777–1855) und Wilhelm Weber (1804–1891). In: MEŸENN, KARL VON (Hrsg.): *Die Großen Physiker. 1. Bd.* München: C. H. Beck Verlag 1997, S. 357–370.

ASSIS, ANDRE KOCH TORRES AND KARL HEINRICH WIEDERKEHR: Weber quoting Maxwell. In: *Mitteilungen der Gauss-Gesellschaft Göttingen*, Nr. 40 (2003), S. 53–74.

WIEDERKEHR, KARL HEINRICH: Kurzbiografie W. Weber. In: *Lexikon der bedeutenden Naturwissenschaftler.* Hrsg. von DIETER HOFFMANN, HUBERT LAITKO, STAFFEN MÜLLER-WILLE unter Mitarbeit von ILSE JAHN. Heidelberg, Berlin: Spektrum Akademischer Verlag 2004, Bd. 3 (Men–Z).

WIEDERKEHR, KARL HEINRICH: Glanzpunkte im Schaffen und Wirken Wilhelm Webers (1804–1891). In: *Mitteilungen der Gauss-Gesellschaft Göttingen*, Nr. 42 (2005), S. 33–42.

WIEDERKEHR, KARL HEINRICH: Wilhelm Weber (1804–1891). In: *Vom Magnetismus zur Elektrodynamik.* Herausgegeben anläßlich des 200. Geburtstages von Wilhelm Weber (1804–1891) und des 150. Todestages von Carl Friedrich Gauß (1777–1855). Katalog zur Ausstellung in der Staatsbibliothek Hamburg, 3. März bis 2. April 2005. Hrsg. von GUDRUN WOLFSCHMIDT. Hamburg: Schwerpunkt Geschichte der Naturwissenschaften, Mathematik und Technik 2005, S. 74–86.

Autors

Prof. Dr. Andre Koch Torres Assis (Campinas, SP, Brazil)

Andre Koch Torres Assis was born in Brazil (1962) and educated at the University of Campinas – UNICAMP, BS (1983), PhD (1987). He spent the academic year of 1988 in England with a post-doctoral position at the Culham Laboratory (United Kingdom Atomic Energy Authority). He spent one year in 1991–1992 as a Visiting Scholar at the Center for Electromagnetics Research of Northeastern University (Boston, USA). From August 2001 to November 2002, and from February to May 2009, he worked at the Institute for the History of Sciences, Hamburg University (Hamburg, Germany) with research fellowships awarded by the *Alexander von Humboldt Foundation* of Germany.

He is the author of *Weber's Electrodynamics* (1994), *Relational Mechanics* (1999), *Inductance and Force Calculations in Electrical Circuits* (with M. A. Bueno, 2001), *The Electric Force of a Current* (with J. A. Hernandes, 2007), and *Archimedes, the Center of Gravity, and the First Law of Mechanics* (2008). He has been professor of physics at UNICAMP since 1989, working on the foundations of electromagnetism, gravitation, and cosmology.

Institute of Physics 'Gleb Wataghin'
University of Campinas—UNICAMP
13083-970 Campinas, SP, Brazil
Homepage: `http://www.ifi.unicamp.br/~assis`
e-mail: `assis@ifi.unicamp.br`

PD Dr. Karl Heinrich Wiederkehr (Hamburg)

Zusammen mit meinem Zwillingsbruder Hans Konrad wurde ich am 1. Februar 1922 in Oftersheim geboren. An der Hebelschule in Schwetzingen (Realgymnasium) machte ich 1941 das Abitur. Ich meldete mich freiwillig zur Kriegsmarine und wurde als Ingenieuroffizier ausgebildet. Nach dem Kriege studierte ich an der Universität Hamburg Physik, Mathematik, Chemie und Philosophie und machte 1949 das Staatsexamen für das Lehramt an Gymnasien. Von 1950 bis 1984 war ich im Hamburger Schuldienst tätig, zuletzt als Studiendirektor und Oberstufenkoordinator. 1962 promovierte ich in Hamburg mit dem von Hans Schimank mir gegebenen Thema über Wilhelm Webers Elektrodynamik.

1967 erschien von mir die Biografie zu Wilhelm Eduard Weber (1804–1891), Bd. 32 der Reihe Große Naturforscher, 1970 erschien gemeinsam mit H.-J. Bersch das Begleitbuch *Klassische Experimente der Physik*, rororo tele (13teilige Sendereihe im NDR). 1974 habilitierte ich mit einem Thema zu René-Just Haüy aus der Kristallographie und erwarb den Status eines Privatdozenten für Geschichte der Naturwissenschaften. Die Habilitationsschrift erschien, aufgegliedert in vier Teilen, in *Centaurus* 1977 und 1978. Für das *Lexikon Große Naturwissenschaftler*, hrsg. von Fritz Krafft, wurden von mir 85 Kurzbiographien geschrieben, für *Die Großen Physiker*, hrsg. von K. v. Meÿenn 1997 (vier längere Biographien).

Des weiteren verfaßte ich zahlreiche Artikel für wissenschaftliche Zeitschriften, hauptsächlich zur Elektrodynamik, Kristallographie und Geophysik.

Bereich Geschichte der Naturwissenschaften
Universität Hamburg
Bundesstrasse 55
D-20146 Hamburg, Germany
Private address: Birkenau 24
D-22087 Hamburg, Germany
e-mail: `Christine_Peters@web.de`

Prof. Dr. Gudrun Wolfschmidt (Hamburg, Germany)

Dissertation *Analysis of close binary systems*, Dr. Remeis Observatory Bamberg, Astronomical Institute of the Friedrich-Alexander University Erlangen-Nuremberg, 1st and 2nd State Examination (physics and mathematics), high school teacher (Gymnasium).

Since 1987 research in history of science in the Deutsches Museum in Munich; conception and realisation of the permanent exhibition "Astronomy and Astrophysics" in the Deutsches Museum (1992, catalogue 1993). From 1992 to 1995 scientific assistant in the research institute for history of science and technology in the Deutsches Museum, different exhibitions (e. g. Copernicus 1994), university teaching and habilitation *Genesis of Astrophysics* (1997) at the Ludwig-Maximilians University in Munich; since 1997 Professor at the Institute for History of Science, Mathematics and Technology of Hamburg University.

Focus of research: History of astronomy and astrophysics (Early Modern Period and $19^{th}/20^{th}$ century) as well as scientific instruments, history of physics, chemistry and technology.

Some book publications/monographs: *Copernicus – Revolutionär wider Willen* (1994), *Milchstraße – Nebel – Galaxien. Strukturen im Kosmos von Herschel bis Hubble* (1995), *Popularisierung der Naturwissenschaften* (2002), *Vom Magnetismus zur Elektrodynamik* (2005), *Development of Solar Research. Entwicklung der Sonnenforschung* (with Axel Wittmann and Hilmar Duerbeck) (2005), *Astronomy in and around Prague* (with Martin Šolc) (2005), *Von Hertz bis Handy* (2007), *Heinrich Hertz (1857–1894) and the Development of Communication* (2008), *Prähistorische Astronomie und Ethnoastronomie* (2008), *"Navigare necesse est" – Geschichte der Navigation* (2008), *Astronomisches Mäzenatentum* (2008), *Hamburgs Geschichte einmal anders – Entwicklung der Naturwissenschaften, Medizin und Technik, Teil 1* (2007) and *Teil 2* (2009), *"Sterne weisen den Weg" – Geschichte der Navigation* (2009), *Cultural Heritage of Astronomical Observatories – From Classical Astronomy to Modern Astrophysics* (2009) and *Astronomie in Nürnberg* (2010); Editor of *Nuncius Hamburgensis, Beiträge zur Geschichte der Naturwissenschaften*.

Institute for History of Science, Hamburg University
Bundesstrasse 55 Geomatikum, D-20146 Hamburg
http://www.math.uni-hamburg.de/home/wolfschmidt/
e-mail: gudrun.wolfschmidt@uni-hamburg.de

Figure 5.1:
Gauß-Weber-Ausstellung des IGN, organisiert von Gudrun Wolfschmidt und Karl Heinrich Wiederkehr, in der Staatsbibliothek Hamburg, 2004/05
Foto: Gudrun Wolfschmidt

List of Figures

0.1	Wilhelm Eduard Weber (1804–1891)	2
0.2	Gauss-Weber-Exhibition *Vom Magnetismus zur Elektrodynamik* of the Institute for History of Science, Mathematics and Technology, Hamburg University, organized by Gudrun Wolfschmidt and Karl Heinrich Wiederkehr, in the State and University Library Hamburg in 2005	9
0.3	Die drei Weber-Brüder: Ernst Heinrich, Wilhelm Eduard und Eduard Friedrich	13
1.1	Wilhelm Eduard Weber (1804–1891) – 1835	16
1.2	Hans Christian Ørsted (1777–1851)	21
1.3	André Marie Ampère (1775–1836)	23
1.4	Michael Faraday (1791–1867)	25
1.5	Weber's conception of Ampère's molecular current. This represents Weber's simplest planetary model of the atom. In this idealized conception, a negative charged particle a follows an elliptical orbit around a positive ponderable electrical mass A. .	44
1.6	Carl Neumann (1832–1925)	71
1.7	The ponderable molecules according to Wilhelm Weber	82
2.1	Wilhelm Eduard Weber (1804–1891) – 1865	102
2.2	Luigi Galvani (1737–1798) und Alessandro Volta (1745–1827) .	104
2.3	James Clerk Maxwell (1831–1879)	107
2.4	Gustav Theodor Fechner (1801–1887)	109
2.5	Tangentenbussole: Das Magnetfeld einer um die Kompaßnadel herumgeführten, vom Strom durchflossenen Kreiswindung lenkt die Nadel ab. Mit diesem Instrument führte Wilhelm Weber die erste absolute elektromagnetische Strommessung durch.	112
2.6	Friedrich Kohlrausch (1840–1910)	115
2.7	Eduard Riecke (1845–1915)	119
2.8	Edwin Herbert Hall (1855–1938)	122
2.9	Paul Drude (1863–1906)	124
2.10	Hendrik Antoon Lorentz (1853–1928), 1916 portraitiert von Menso Kamerlingh Onnes (1860–1925)	129
2.11	Niels Bohr (1885–1962)	131

2.12	Richard Chace Tolman and Albert Einstein, 1932	134
2.13	Arnold Sommerfeld (1868–1951) – 1897	136
2.14	Enrico Fermi (1901–1954), 1943/49	137
2.15	Paul Adrian Maurice Dirac (1902–1984)	138
2.16	Felix Bloch (1905–1983)	141
3.1	Wilhelm Eduard Weber (1804–1891)	150
3.2	Johann Karl Friedrich Zöllner (1834–1882)	162
3.3	Annalen der Physik (4. Serie), Band 306 (1900), ed. by Paul Drude	163
5.1	Gauß-Weber-Ausstellung des IGN, organisiert von Gudrun Wolfschmidt und Karl Heinrich Wiederkehr, in der Staatsbibliothek Hamburg, 2004/05	170
5.2	Gauß-Weber-Telegraph 1833	178

Nuncius Hamburgensis
Beiträge zur Geschichte der Naturwissenschaften

Norderstedt: Books on Demand (nur Bd. 2, 6, 7, 8, 10, 11, 14 und 15)

Hamburg: tredition Verlag **tredition** science (alle anderen Bände).

Hg. von Gudrun Wolfschmidt,
Bereich Geschichte der Naturwissenschaften, Fachbereich Mathematik,
Fakultät für Mathematik, Informatik und Naturwissenschaften (MIN),
Universität Hamburg – ISSN 1610-6164

*Diese Reihe „Nuncius Hamburgensis" wird gefördert von
der Hans Schimank-Gedächtnisstiftung. Dieser Titel wurde inspiriert
von „Sidereus Nuncius" und von „Wandsbeker Bote".*

  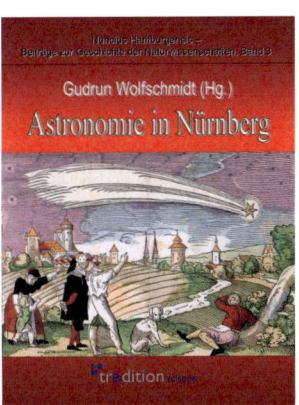

- Band 1 (2009):
  Hans Schimank (1888–1979) Ausgewählte Schriften.
  Mit einem Beitrag ‚Hans Schimanks Otto von Guericke' von Fritz Krafft.
  Bearbeitet von Timo Engels und Igor Abdrakhmanov.

- Band 2 (2007):
  Wolfschmidt, Gudrun (Hg.): *Hamburgs Geschichte einmal anders –
  Entwicklung der Naturwissenschaften, Medizin und Technik – Teil 1.*

- Band 3 (2010):
  Wolfschmidt, Gudrun (Hg.): *Astronomie in Nürnberg*.
  Proceedings der Tagung vom 2.–3. April 2005 in Nürnberg anläßlich des 500. Todestages von Bernhard Walther (1430–1504) und des 300. Todestages von Georg Christoph Eimmart (1638–1705).

- Band 4 (2011):
  Wolfschmidt, Gudrun (Hg.): *Entwicklung der Astrophysik*.
  Proceedings des Kolloquiums des Arbeitskreises Astronomiegeschichte in der Astronomischen Gesellschaft am 26. September 2005 in Köln.

- Band 5 (2011):
  Wolfschmidt, Gudrun (Hg.):
  Anfänge der Theoretischen Physik in Hamburg.
  Vorwort von Kurt Scharnberg und Klaus Fredenhagen.

- Band 6 (2007):
  Wolfschmidt, Gudrun (Hg.): *Von Hertz zum Handy – Entwicklung der Kommunikation*. Begleitbuch zur Ausstellung zum 150. Geburtstag von Heinrich Hertz (1857–1894).

- Band 7 (2009):
  Wolfschmidt, Gudrun (Hg.): *Hamburgs Geschichte einmal anders – Entwicklung der Naturwissenschaften, Medizin und Technik, Teil 2*.

- Band 8 (2008):
  Wolfschmidt, Gudrun (Hg.):
  Prähistorische Astronomie und Ethnoastronomie.
  Proceedings des Kolloquiums des Arbeitskreises Astronomiegeschichte
  in der Astronomischen Gesellschaft am 24. September 2007 in Würzburg.

- Band 9 (2012):
  Wolfschmidt, Gudrun (Hg.):
  Naturwissenschaft, Technik und Kultur in London.

- Band 10 (2008):
  Wolfschmidt, Gudrun (ed.): *Heinrich Hertz (1857–1894)
  and the Development of Communication.* Proceedings of the
  International Scientific Symposium in Hamburg, Oct., 8–12, 2007.

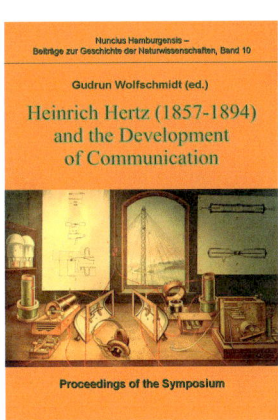

- Band 11 (2008):
  Wolfschmidt, Gudrun (Hg.):
  Astronomisches Mäzenatentum.
  Proceedings des Symposiums in der Kuffner-Sternwarte in Wien,
  „Astronomisches Mäzenatentum in Europa", 7.–9. Oktober 2004.

- Band 12 (2011):
  Wolfschmidt, Gudrun (Hg.):
  Astronomie in neuen Wellenlängen – Astronomy in New Wavelength.
  Proceedings des Kolloquiums des Arbeitskreises Astronomiegeschichte
  in der Astronomischen Gesellschaft am 24. September 2007 in Würzburg.

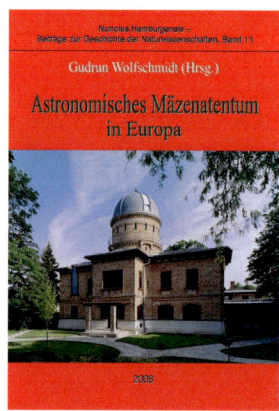  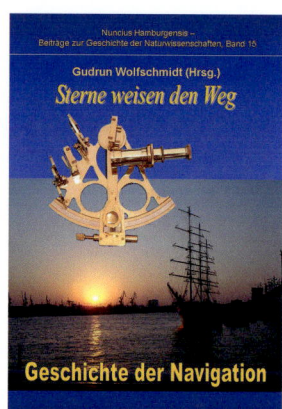

- Band 13 (2011):
  Cura, Katrin: *Alchemie im Deutschen Museum.*
  Bearbeitet von Gudrun Wolfschmidt.

- Band 14 (2008):
  Wolfschmidt, Gudrun (Hg.):
  „Navigare necesse est" – Geschichte der Navigation.
  Begleitbuch zur Ausstellung 2008/09 in Hamburg und Nürnberg.

- Band 15 (2009):
  Wolfschmidt, Gudrun:
  „Sterne weisen den Weg" – Geschichte der Navigation.
  Katalog zur Ausstellung 2008/10 in Hamburg und Nürnberg.

- Band 16 (2011):
  Wolfschmidt, Gudrun (Hg.):
  *Simon Marius, der fränkische Galilei,
  und die Entwicklung des astronomischen Weltbildes.*

- Band 17 (2011):
  Cura, Katrin:
  Auf den Leim gehen – Geschichte der Klebstoffe.
  Hg. von Gudrun Wolfschmidt.

- Band 18 (2011):
  Wolfschmidt, Gudrun (Hg.):
  Farben in Kulturgeschichte und Naturwissenschaft.
  Begleitbuch zur Ausstellung in Hamburg 2010–2012.

- Band 19 (2011):
  Andre Koch Torres Assis und Karl Heinrich Wiederkehr
  und Gudrun Wolfschmidt:
  Weber's Planetary Model of the Atom.
  Ed. by Gudrun Wolfschmidt.

- Band 20 (2011):
  Wolfschmidt, Gudrun (Hg.):
  *Hamburgs Geschichte einmal anders –
  Entwicklung der Naturwissenschaften,
  Medizin und Technik, Teil 3.*

- Band 21 (2012):
  Wolfschmidt, Gudrun (Hg.):
  Vom Abakus zum Computer – Geschichte der Rechentechnik.
  Begleitbuch zur Ausstellung in Hamburg.

- Band 22 (2011):
  Wolfschmidt, Gudrun (ed.):
  Colours in Culture and Science. 200 Years Goethe's Colour Theory.
  Proceedings of the Interdisciplinary Symposium in Hamburg,
  October 12–15, 2010.

Web-Seite zur aktuellen Information

```
http://www.math.uni-hamburg.de/spag/ign/research/nuncius.htm
```

Figure 5.2:
Gauß-Weber-Telegraph 1833
Deutsches Museum München (BN 35826)

Index

A

Aharonov, Yakir (*1932), 157
Ampère, André Marie (1775–1836), 14, 18, 19, 22–24, 26, 27, 31, 33–36, 41, 42, 44–47, 51, 53, 63, 64, 66–68, 71, 72, 74, 91–94, 96, 103–106, 110, 111, 130, 142, 143, 154–156, 164
Arago, François Dominique (1786–1853), 24

B

Banks, Joseph (1743–1820), 96
Becquerel, Alexandre Edmond (1820–1891), 62
Becquerel, Henri (1852–1908), 62
Beetz, Wilhelm von (1822–1886), 114
Benedicks, Carl (fl. 1929), 133, 142
Berthollet, Claude Louis (1748–1822), 24
Berzelius, Jöns Jacob (1779–1848), 24, 87
Besemann, Friedrich (1796–1854), 4
Bethe, Hans Albrecht (1906–2005), 10, 137, 139, 147
Biot, Jean-Baptiste (1774–1862), 19, 24, 92, 93
Bloch, Felix (1905–1983), 139–142

Bohm, David Joseph (1917–1992), 157, 161
Bohr, Niels (1885–1962), 62, 84, 127, 130–133, 142, 146, 147
Boltzmann, Ludwig (1844–1906), 126, 127, 135, 137, 139
Born, Max (1882–1970), 10
Bose, Satyendra Nath (1894–1974), 135
Broglie, Louis-Victor de (1892–1987), 138, 140, 157, 158
Bunsen, Robert Wilhelm (1811–1899), 75

C

Casimir, Hendrik Brugt Gerhard (1909–2000), 10
Cavendish, Henry (1731–1810), 28
Clausius, Rudolf (1822–1888), 126
Copernicus, Nikolaus (1473–1543), 169
Coulomb, Charles Augustin (1736–1806), 27, 53, 60–62, 85, 87

D

Davy, Humphry (1778–1829), 24, 87
Debye, Peter (1884–1966), 132, 140, 142, 145

Dirac, Paul Adrien Maurice (1902–1984), 10, 135, 138
Drude, Paul (1863–1906), 66, 92, 93, 103, 117, 118, 121, 124–128, 130, 132, 133, 139, 143, 144, 163
Du Bois-Reymond, Emil (1818–1896), 15
Dubois-Reymond, Estelle (1865–1955), 15
Dulong, Pierre Louis (1785–1838), 132

E

Eimmart, Georg Christoph (1638–1705), 174
Einstein, Albert (1879–1955), 92, 94, 134, 135, 161
Ettingshausen, Andreas Freiherr von (1796–1878), 99, 123, 149
Everdingen, E. van (fl. 1897), 118

F

Faraday, Michael (1791–1867), 19, 24–31, 41, 69, 70, 72, 93, 96, 99, 103–108, 110, 113, 116, 117, 143, 148, 149
Fechner, Gustav Theodor (1801–1887), 19, 31, 92–94, 106, 108–111, 143
Fermi, Enrico (1901–1954), 135, 137, 147
Franz, Rudolph (1827–1902), 63, 115, 116, 120, 126, 127, 132, 139, 148
Fresnel, Augustin (1788–1827), 35, 36, 93

G

Galilei, Galileo (1564–1642), 156, 176
Galvani, Luigi (1737–1798), 19, 93, 98, 99, 104, 117, 146, 159, 160
Gans, Richard [Ricardo] (1880–1954), 123, 143
Gauß, Carl Friedrich (1777–1855), 8, 9, 12, 14, 15, 92, 94, 109, 148, 149, 157, 158, 161, 164–166, 170, 178
Geiger, Hans (1882–1945), 131, 137, 143, 147
Giese, Wilhelm (fl. 1889), 117, 143
Graßmann, Hermann Günther (1809–1877), 155, 156
Grimsehl, Ernst (1861–1914), 132, 147
Guericke, Otto von (1602–1686), 173

H

Hall, Edwin Herbert (1855–1938), 122, 123, 130, 142
Hankel, Wilhelm Gottlieb (1814–1899), 89
Haüy, René-Just (1743–1822), 9, 168
Heisenberg, Werner (1901–1976), 10, 139, 142, 144
Helmholtz, Hermann von (1821–1894), 14, 15, 88, 117, 156
Herschel, Friedrich Wilhelm (1738–1822), 169
Hertz, Heinrich (1857–1894), 10, 88, 89, 100, 106, 110, 116, 117, 125, 126, 169, 174, 175
Hoppe, Edmund (1854–1928), 14
Hubble, Edwin Powell (1889–1953), 169
Humboldt, Alexander von (1769–1859), 14, 90

I

J

Joos, Georg (1894–1959), 127, 132, 144
Jordan, Ernst Pascual (1902–1980), 10
Joule, James Prescott (1818–1889), 62, 116

K

Kalischer, Salomon (1845–1924), 105, 143
Kamerlingh Onnes, Menso (1860–1925), 129
Kepler, Johannes (1571–1630), 20, 43, 44, 49, 50, 61–63, 77
Kirchhoff, Gustav Robert (1824–1887), 75, 154
Kohlrausch, Friedrich Wilhelm Georg (1840–1910), 4, 90, 94, 100, 102, 114–116, 120, 121, 144
Kohlrausch, Rudolf (1809–1858), 32, 54, 90, 94, 99, 100, 114, 149, 158, 161, 164

L

Langevin, Paul (1872–1946), 130, 144
Leibniz, Gottfried Wilhelm (1646–1716), 159
Lenz, Heinrich Friedrich Emil (1804–1865), 62
Lewis, Gilbert Newton (1875–1946), 87
Lindemann, Frederick Alexander (1886–1957), 132, 133, 144

Lorentz, Hendrik Antoon (1853–1928), 14, 103, 113, 117, 118, 123, 127–130, 132, 133, 138, 139, 144, 145, 155, 160
Lorenz, Ludwig (1829–1891), 63, 116, 120, 126, 127, 132, 139
Ludwig, Carl (1816–1895), 15
Ludwig, Karl → Ludwig, Carl, 15

M

Mach, Ernst (1838–1916), 92, 93, 95, 154, 155, 157–161
Marius, Simon (1573–1624), 176
Maxwell, James Clerk (1831–1879), 14, 15, 19, 24, 28–31, 92, 94, 99, 100, 103, 106–108, 110, 111, 116, 117, 125–128, 133, 135, 137, 144, 145, 149, 158, 160, 166
Meyer-Abich, Adolf (1895–1971), 9, 11, 165
Mossotti, Ottaviano Fabrizio (1791–1863), 77, 88, 94

N

Nernst, Walther (1864–1941), 123, 135
Neumann, Carl G. (1832–1925), 71–74, 95, 113, 114, 145
Neumann, Franz Ernst (1798–1895), 71, 113, 145
Newton, Isaac (1642–1726 a.St./1643–1727), 27, 51, 53, 61, 62, 70, 77, 95, 100, 155

O

Ørsted, Hans Christian (1777–1851), 19–22, 24, 26, 27, 34, 36,

51, 95, 99, 104–106, 146, 148
Ohm, Georg Simon (1789–1854), 19, 36, 41, 42, 47, 51, 64, 66, 94, 95, 111, 120, 164

P

Pauli, Wolfgang (1900–1958), 10, 135, 137, 140, 144, 146
Peierls, Rudolf Ernst (1907–1995), 139, 142, 146
Peltier, Jean Charles Athanase (1785–1845), 114, 121, 146
Petit, Alexis Thérèse (1791–1820), 132
Planck, Max (1858–1947), 126, 127, 135, 139, 146
Poggendorff, Johann Christian (1796–1877), 15, 100, 110, 113, 143, 144, 148
Pohl, Robert Wichard (1884–1976), 120
Poisson, Siméon-Denis (1781–1840), 28

Q

R

Reissner, Hans Jacob (1874–1967), 160
Richardson, Owen Willans (1879–1959), 139
Riecke, Eduard (1845–1915), 15, 44, 88, 95, 98, 100, 103, 117–121, 125–128, 130, 132–134, 146, 147
Rutherford, Ernest (1871–1937), 62, 84, 131

S

Sale, R. E. [Lieut.] (fl. 1873), 89
Savart, Félix (1791–1841), 24
Scheel, Karl (1866–1936), 137, 143, 147
Schimank, Hans (1888–1979), 11, 12, 100, 168, 173
Schrödinger, Erwin (1887–1961), 10, 95, 134, 138, 160, 161
Schweigger, Johann Salomo Christoph (1779–1857), 24, 95
Seebeck, Thomas Johann (1770–1831), 114, 147
Seegers, Carolus (fl. 1864), 96
Seeliger, Rudolf (1886–1965), 121, 123, 127, 130, 133, 135, 140, 147
Serres, Marcel de (1780–1862), 106
Servus, Arminius (∗1858), 96
Sommerfeld, Arnold (1868–1951), 10, 103, 133–140, 142, 143, 147
Stark, Johannes (1874–1957), 140, 145–147
Stewart, Thomas Dale (1890–1958), 96, 108, 133, 134, 147
Sticker, Bernhard (1906–1977), 8, 12
Stoney, George Johnstone (1826–1911), 118

T

Thomson, Joseph John (1856–1940), 15, 118, 131, 133, 147
Thomson, William [Lord Kelvin] (1824–1907), 116, 121, 147
Tisserand, François Félix (1845–1896), 96
Tolman, Richard Chace (1881–1948), 31, 96, 108, 133, 134, 147

Tomaschek, Rudolf (1895–1966), 132, 147

U

Voigt, Woldemar (1850–1919), 98, 118, 119, 125, 147
Volta, Alessandro (1745–1827), 20, 24, 95, 96, 100, 104, 158, 164

V

W

Walther, Bernhard (1430–1504), 174
Weber, Eduard Friedrich (1806–1871), 12, 13, 15, 110, 149, 165
Weber, Ernst Heinrich (1795–1878), 12, 13, 15, 18, 70, 100, 110, 149, 164, 165
Weber, Heinrich (1839–1928), 94, 97–100, 159, 160, 162
Weber, Wilhelm Eduard (1804–1891), 1–5, 8, 9, 11–19, 31–34, 36, 41–49, 51, 53–90, 92, 94–103, 106, 108–114, 118, 126, 130, 148–161, 164–166, 168, 170, 177, 178
Wiedemann, Eilhard (1852–1928), 118, 146
Wiedemann, Gustav Heinrich (1826–1899), 63, 115–118, 120, 126, 127, 132, 139, 143, 146, 148
Wien, Wilhelm (1864–1928), 135, 137, 149
Wilson, Alan Herries (1906–1995), 140
Wollaston, William Hyde (1766–1828), 24

X

Y

Z

Zeeman, Pieter (1865–1943), 118
Zöllner, Johann Karl Friedrich (1834–1882), 75, 77, 89, 94, 100, 101, 162

www.tredition.de

Über tredition

Der tredition Verlag wurde 2007 in Hamburg gegründet und ermöglicht Autoren das Publizieren von e-Books, audio-Books und print-Books. Autoren veröffentlichen ihre Bücher selbständig oder auf Wunsch mit der Unterstützung von tredition. print-Books sind in allen Buchhandlungen sowie bei Online-Händlern gedruckter Bücher erhältlich. e-Books und audio-Books können auf Wunsch der Autoren neben dem tredition Web-Shop auch bei weiteren führenden Online-Portalen zum Verkauf angeboten werden.

Auf www.tredition.de veröffentlichen Autoren in wenigen leichten Schritten ihr Buch. Zusätzlich bieten zahlreiche Literatur- Partner (das sind Lektoren, Übersetzer, Hörbuchsprecher und Illustratoren) ihre Dienstleistung an, um Manuskripte zu verbessern oder die Vielfalt zu erhöhen. Autoren können dieses Angebot nutzen und vereinbaren unabhängig von tredition mit Literatur- Partnern ihre Zusammenarbeit und partizipieren gemeinsam am Erfolg des Buches.